Pesticides in the Atmosphere

Distribution, Trends, and Governing Factors

Michael S. Majewski, U.S. Geological Survey, Sacramento, California
Paul D. Capel, U.S. Geological Survey, St. Paul, Minnesota

Volume One of the Series
Pesticides in the Hydrologic System

Robert J. Gilliom, Series Editor
U.S. Geological Survey
National Water Quality Assessment Program

CRC Press
Taylor & Francis Group
Boca Raton London New York

CRC Press is an imprint of the
Taylor & Francis Group, an **informa** business

First published by Ann Arbor Press, Inc.

Published 2019 by CRC Press
Taylor & Francis Group
6000 Broken Sound Parkway NW, Suite 300
Boca Raton, FL 33487-2742

First issued in paperback 2020

ISBN 13: 978-0-367-57965-4 (pbk)
ISBN 13: 978-1-57504-004-2 (hbk)

Visit the Taylor & Francis Web site at
http://www.taylorandfrancis.com

and the CRC Press Web site at
http://www.crcpress.com

Library of Congress Cataloging-in-Publication Data

Majewski, Michael S.
    Pesticides in the atmosphere : distribution, trends, and governing
    factors / Michael S. Majewski, Paul D. Capel
        p.   cm. -- (Pesticides in the hydrologic system : v. 1)
        Includes bibliographical references (p.   ) and index.
        ISBN 1-57504-004-2
        1. Pesticides--Environmental aspects--United States. 2. Air-
-Pollution--United States. 3. Atmospheric diffusion--United States.
I. Capel, Paul D.   II. Title.   III. Series.
TD887.P45M35   1995
628.5'3--dc20   95-34078

Library of Congress Card Number 95-34078

INTRODUCTION TO THE SERIES

Pesticides in the Hydrologic System is a series of comprehensive reviews and analyses of our current knowledge and understanding of pesticides in the water resources of the United States and of the principal factors that influence contamination and transport. The series is presented according to major components of the hydrologic system--the atmosphere, surface water, bed sediments and aquatic organisms, and ground water. Each volume:

- summarizes previous review efforts;

- presents a comprehensive tabulation, review, and analysis of studies that have measured pesticides and their transformation products in the environment;

- maps locations of studies reviewed, with cross references to original publications;

- analyzes national and regional patterns of pesticide occurrence in relation to such factors as the use of pesticides and their chemical characteristics;

- summarizes processes that govern the sources, transport, and fate of pesticides in each component of the hydrologic system;

- synthesizes findings from studies reviewed to address key questions about pesticides in the hydrologic system, such as:

  How do agricultural and urban areas compare?

  What are the effects of agricultural management practices?

  What is the influence of climate and other natural factors?

  How do the chemical and physical properties of a pesticide influence its behavior in the hydrologic system?

  How have past study designs and methods affected our present understanding?

  Are water-quality criteria for human health or aquatic life being exceeded?

  Are long-term trends evident in pesticide concentrations in the hydrologic system?

This series is unique in its focus on review and interpretation of reported direct measurements of pesticides in the environment. Each volume characterizes hundreds of studies conducted during the past four decades. Detailed summary tables include such features as spatial and temporal domain studied, target analytes, detection limits, and compounds detected for each study reviewed.

Pesticides in the Hydrologic System is designed for use by a wide range of readers in the environmental sciences. The analysis of national and regional patterns of pesticide occurrence, and their relation to use and other factors that influence pesticides in the hydrologic system, provides a synthesis of current knowledge for scientists, engineers, managers, and policy makers at all levels of government, in industry and agriculture, and in other organizations. The interpretive analyses and summaries are designed to facilitate comparisons of past findings to current and future findings. Data of a specific nature can be located for any particular area of the country. For educational needs, teachers and students can readily identify example data sets that meet their requirements. Through its focus on the United States, the series covers a large portion of the global database on pesticides in the hydrologic system and international readers will find much that applies to other areas of the world. Overall, the goal of the series is to provide readers from a broad range of backgrounds in the environmental sciences with a synthesis of the factual data and interpretive findings on pesticides in the hydrologic system.

The series has been developed as part of the National Water-Quality Assessment Program of the U. S. Geological Survey, Department of Interior. Assessment of pesticides in the nation's water resources is one of the top priorities for the Program, which began in 1991. This comprehensive national review of existing information serves as the basis for design and interpretation of studies of pesticides in major hydrologic systems of the United States now being conducted as part of the National Water-Quality Assessment.

Series Editor

Robert J. Gilliom
U. S. Geological Survey

PREFACE

Most people are aware of and concerned with the health effects of pesticide residues in the water they drink and the food they eat, but many are surprised to learn that pesticides are commonly found in air and rain. Scientific studies of pesticides in various atmospheric matrices (air, rain, snow, aerosols, and fog) have been ongoing for 40 years. When taken together, these studies, many of which are small and focused, provide a significant contribution to answering the questions when, where, how, and why pesticides are in the atmosphere. The studies also make an important contribution to our understanding of the environmental effects of pesticides, particularly on water quality. The broader scientific and political communities, though familiar with the impact of pesticides on water quality, are relatively unaware of the significance of the contribution of atmospheric transportation and deposition of pesticides to water quality. In retrospect, the effects of DDT on the bald eagle, first described by Rachel Carson, may have been largely caused by the atmospheric distribution of pesticides.

This book was written with the goal of building upon the foundation of what we presently know about pesticides in the atmosphere to better understand their effects in the hydrologic system. To accomplish this, we have compiled and evaluated most of the published studies that have investigated the occurrences and behavior of pesticides in the atmosphere; synthesized the varied information from these studies to characterize the common threads and main conclusions; and identified major needs for improved understanding of pesticides in the atmosphere and the significance to water quality. As such, this book is intended to serve as a resource, text, and reference to a wide spectrum of scientists, students, and water managers, ranging from those primarily interested in the extensive compilations of references, to those looking for interpretive analyses and conclusions. For those not familiar with the studies of pesticides in the atmosphere, it can serve as a comprehensive introduction.

As part of the review and interpretation, it was necessary to include brief reviews of how pesticides enter the atmosphere, how meteorology influences their behavior and transport, and how airborne pesticides are redeposited to terrestrial and aquatic ecosystems. These brief reviews of environmental processes summarize the research findings of the scientific literature. Although some aspects of pesticide behavior in the environment are well understood, many aspects have a distinct lack of data that limits our understanding.

This book was made possible by the National Water-Quality Assessment Program and the foresight and commitment of its leadership team and the U.S. Geological Survey to understand the behavior and transport of pesticides in all aspects of the hydrologic cycle. We are greatly indebted to Loreen Kleinschmidt of the Toxicology Documentation Center at the University of California, Davis for her tireless support in conducting literature searches, obtaining many of the references, and assisting in many other ways during the research and writing phase of this book. Naomi Nakagaki produced many of the maps and patiently tolerated our countless updates. Tom Sklarsky and Susan Davis provided excellent and conscientious editing and manuscript preparation. We also thank Donald A. Goolsby and William T. Foreman for thorough reviews of this book. Both made many excellent suggestions that greatly improved the quality of the final product.

Michael S. Majewski
Paul D. Capel

EDITOR'S NOTE

This work originally was prepared as a United States Geological Survey report. Though the report has been edited for commercial publication, some of the style and usage incorporated is based on the United States Geological Survey publication guidelines (Suggestions to Authors, 7th ed., 1991). For example, references with more than two authors cited within the text are written as "Smith and others (19xx)," rather than "Smith, et al. (19xx)," and some common use compound adjectives are hyphenated when used as a modifier (e.g., ground-water supply and surface-water supply). For units of measure, the metric system is used except for the reporting of pesticide use. When quoting from other sources, the original system is used. Some of the longer tables are placed at the end of the chapter to maintain less disruption of text.

CONTENTS

FIGURES

LIST OF TABLES

CONVERSION FACTORS

Multiply	By	To obtain
centimeter (cm)	0.3937	inch
cubic meter (m^3)	35.31	cubic foot
gram (g)	0.03527	ounce, avoirdupois
hectare (ha)	2.469	acre
kilogram (kg)	2.205	pound, avoirdupois
pound, avoirdupois (lb)	0.4536	kilogram
kilometer (km)	0.6214	mile
liter (L)	0.2642	gallon
meter (m)	3.281	foot
square kilometer (km^2)	0.3861	square mile
square meter (m^2)	10.76	square foot

Temperature is given in degrees Celsius (°C), which can be converted to degrees Fahrenheit (°F) by the following equation:

$$°F = 1.8(°C) + 32$$

ABBREVIATIONS

α, alpha
β, beta
δ, delta
γ, gamma
g/ha/d, gram per hectare per day
lb a.i., pounds active ingredient
lb a.i./yr, pounds active ingredient per year
kg/yr, kilogram per year
ng, nanogram
ng/L, nanogram per liter
ng/m^3, nanogram per cubic meter
ng/smpl, nanogram per sample
μg, microgram
μg/g, microgram per gram
μg/L, microgram per liter
$\mu g/m^2/yr$, microgram per square meter per year
μg/smpl, microgram per sample
μm, micrometer
L/ha, liter per hectare
mg/ha, milligram per hectare
mg/kg, milligram per kilogram
mm, millimeter
nm, nanometer
Pa, pascal
$Pa-m^3/mole$, Pascal cubic meter per mole
pg, picogram
pg/m^3, picogram per cubic meter

ACRONYMS

Dep, deposition
H, Henry's law (in values of $Pa-m^3/mole$)
<LOD, less than analytical limit of detection
MCL, maximum contaminant level
NAS, National Academy of Sciences
NAWQA, National Water Quality Assessment
NADP/NTN, National Atmospheric Deposition Program/National Trends Network
ND, not detected
NR, not reported
nsg, no standard or guideline exists for this compound
PAHs, polycyclic aromatic hydrocarbons
PCBs, polychlorinated biphenyls
OA, oxygen analog transformation of the parent compound
OSHA, Occupational Safety and Health Administration
ptcl, particulate matter
ppm, parts per million
TWA, time-weighted average
USEPA, U.S. Environmental Protection Agency

PESTICIDES IN THE ATMOSPHERE

Distribution, Trends, and Governing Factors

Michael S. Majewski and Paul D. Capel

ABSTRACT

A comprehensive review of existing literature on the occurrence and distribution of pesticides in the atmosphere of the United States and adjoining Canadian provinces showed that the atmosphere is an important part of the hydrologic cycle that acts to distribute and deposit pesticides in areas far removed from their application sites. A compilation of existing data shows that pesticides have been detected in the atmosphere throughout the nation. Most of the available information on pesticides in the atmosphere is from small-scale, short-term studies that seldom lasted more than one year. Only two national-scale, multiyear studies were done since the late 1960's that analyzed for a wide variety of pesticides in air that were in current use at the time. Another large-scale study was done during 1990-91, but was limited to the midwestern and northeastern United States and only analyzed for two classes of herbicides in wet deposition. Most of the pesticides analyzed for were detected in either air or rain, and represent about 25 percent of the total number of insecticides, herbicides, and fungicides in current use. The geographical distribution of studies, and the type of sampling and analysis were highly variable with most of the historical study efforts concentrated in the Great Lakes area and California. Air and rain were the main atmospheric matrices sampled, but pesticides were also detected in fog and snow.

Reported pesticide concentrations in air and rain were frequently positively correlated to their regional agricultural use. Deviations from this relation could usually be explained by nonagricultural use of pesticides, sampling and analytical difficulties, and environmental persistence. High concentrations of locally used pesticides were found to occur seasonally, usually in conjunction with spring planting of row crops and warm temperatures, but high concentrations also occurred during winter months in those areas where dormant orchards were sprayed. The environmentally more persistent pesticides were detected in the atmosphere at low concentrations throughout the year.

Deposition of airborne pesticides can have significant effects on water quality, but neither the nature of nor the magnitude of these effects can be determined with certainty on the basis of the type of data currently available. The lack of consistent, long-term regional and national monitoring and study of pesticides in atmospheric matrices severely limits assessment capability.

CHAPTER 1

Introduction

About 1.1 billion pounds of pesticides currently are used each year in the United States to control many different types of weeds, insects, and other pests in a wide variety of agricultural and nonagricultural settings as shown in Figure 1.1 (Aspelin and others, 1992; Aspelin, 1994). Total use and the number of different chemicals applied have grown steadily since the early 1960's, when the first reliable records were kept. For example, national use of herbicides and insecticides on cropland and pasture grew from 190 million lb a.i. in 1964 to 560 million pounds in 1982 (Gilliom and others, 1985), and was estimated to be about 630 million pounds in 1988 (Gianessi and Puffer, 1990, 1992a,b). Increased use has resulted in increased crop production, lower maintenance costs, and control of public health hazards. In addition, however, concerns about the potential adverse effects of pesticides on the environment and human health have grown steadily.

In many respects, the greatest potential for unintended adverse effects of pesticides is through contamination of the hydrologic system, which supports aquatic life and related food chains and is used for recreation, drinking water, and many other purposes. Water is one of the primary mechanisms by which pesticides are transported from targeted application areas to other parts of the environment and, thus, there is potential for movement into and through all components of the hydrologic cycle (see Figure 1.2).

The atmosphere is an important component of the hydrologic cycle to consider in assessing the effect of pesticides in the environment. Pesticides have been recognized as potential air pollutants since 1946 (Daines, 1952). Early in the history of agricultural pesticide use, off-target drift of the applied pesticides was a concern, and much effort has been expended studying the factors that affect drift and the best ways to control it (Akesson and Yates, 1964; Yates and Akesson, 1973). On the other hand, mosquito abatement and other large-scale programs to eradicate such pests as the Mediterranean fruit fly and the Japanese beetle, are examples of pesticide applications directly into the atmosphere with the intention of maximizing the coverage area using aerial drift.

Until the 1960's, atmospheric pesticide contamination was generally thought of as a "local" problem caused by spray drift. Long-range movement of pesticides was thought to be minimal, if any, because of their physical and chemical properties (low volatility and low solubility in water). The detection of DDT (see glossary for chemical names of pesticides) and other organochlorine compounds in fish and mammals in the Arctic (Cade and others, 1968, Addison and Smith, 1974) and Antarctic (George and Frear, 1966; Sladen and others, 1966; Peterle, 1969) changed this notion. These organochlorine residues could, in some cases, be

3

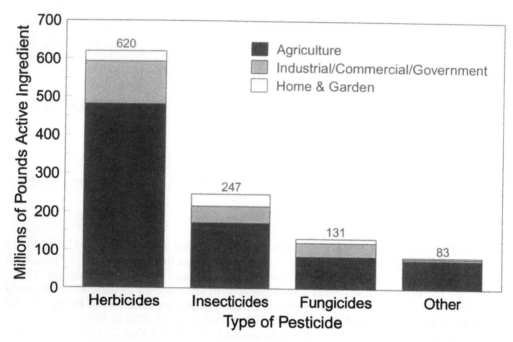

FIGURE 1.1. Estimated mass of total pesticides used in the United States during 1993 for agriculture, industrial/commercial/government, and home and garden (from Aspelin, 1994).

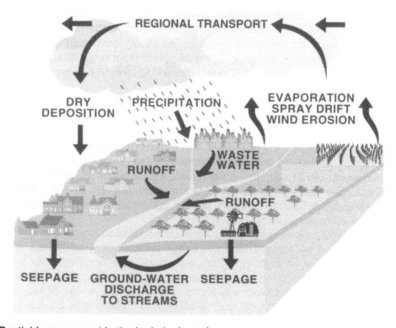

FIGURE 1.2. Pesticide movement in the hydrologic cycle.

attributed to movement in surface water and the distribution through the food chain, but atmospheric deposition had to be considered as a main source of this contamination in many remote areas. The atmosphere is now recognized as a major pathway by which pesticides and other organic and inorganic compounds are transported and deposited in areas sometimes far removed from their sources.

1.1 PURPOSE

This book presents the results of a review of the current understanding of pesticides in the atmosphere (that includes air, rain, snow, fog, and aerosols) of the United States, with an emphasis on the integration and analysis of information from studies across a wide range of spatial and temporal scales. The objectives of the review were to evaluate and assess, to the degree possible from existing information, the occurrence and distribution of pesticides in atmospheric compartments, factors that affect their concentrations and movement in the atmosphere, and the potential significance that pesticides in the atmosphere pose to water quality. In addition, future study needs are addressed on the basis of conclusions from the review. This review of pesticides in the atmosphere is one in a series of reviews of current knowledge of pesticide contamination of the hydrologic system, which is being done as part of the Pesticide National Synthesis project of the U.S. Geological Survey, National Water-Quality Assessment (NAWQA) program. Other reviews in the series focus on pesticides in surface water, ground water, and streambed sediment and aquatic biological tissues. These national topical reviews of published studies on pesticides are intended to complement more detailed studies done in each individual NAWQA study unit; for example, major hydrologic basins, typically 10,000 to 30,000 mi^2 (Hirsch and others, 1988; Gilliom and others, 1995).

1.2 PREVIOUS REVIEWS

Several reviews of existing information have been published on one or more specific aspects of pesticides in the atmosphere, such as the depositional processes, vapor-particle partitioning, some of the predominant reactions, and transport mechanisms. Table 1.1 lists these reviews and briefly describes their scope. Most of these reviews focus on a particular class of pesticide, or include selected pesticides as a subset of a larger group of organic pollutants. Many of the discussions are primarily theoretical. Several of the reviews in Table 1.1 deal with organochlorine pesticides, many of which are no longer used in the United States and Canada. Others discuss in detail environmental processes that can affect airborne organic compounds, but do not specifically address pesticides. Together, these reports provide a relatively complete overview of the range of factors that affect the sources, transport, and fate of pesticides in the atmosphere, but, except for organochlorine pesticides in the Great Lakes region, do not provide a broad perspective on the occurrence, distribution, and significance of pesticides in the atmosphere of the United States.

1.3 APPROACH

This review focuses primarily on studies of pesticides in the atmosphere of the United States. Studies from outside the United States, mainly Canada, and laboratory and process studies were selectively reviewed to help explain particular phenomena or occurrences. The goal of the review process was to locate all significant studies within this scope that have been published in an accessible report format, including journal articles, federal and state reports, and university report series. The studies reviewed were assembled through use of bibliographic data searches (National Technical Information Service and ChemAbstracts), personal collections, and bibliographies from reviewed reports. Studies at all spatial scales, from individual sites or fields to multistate regional studies, were included.

The studies were evaluated and are presented in four primary phases. First, all studies reviewed are tabulated along with selected study features such as location, spatial scale, timeframe, number of sites, media sampled, and target analytes. This serves as an overview and reference to the studies reviewed and provides the basis for an initial characterization of the nature, degree, and emphasis of study effort.

Second, a national perspective on the occurrence and geographical distribution of pesticides in the atmosphere is developed from the observations reported in the reviewed studies, with particular emphasis on the large-scale studies. This overview defines the geographic nature of the issue for different pesticides and different atmospheric media. Although limited by the biases inherent in the studies reviewed, it provides a perspective on the degree to which atmospheric contamination of pesticides may be a problem and on what some of the basic priorities are.

The third phase of the approach is a summary review of the primary factors that affect pesticide concentrations in the atmosphere. This provides a basis for understanding observed patterns in occurrence and distribution and for posing and addressing more refined questions.

The fourth part of the review is a detailed analysis of what existing information can tell us about the answers to specific questions concerning pesticides in the atmosphere. The questions were developed to reflect the range of basic factors that need to be understood in order to evaluate the causes, degree, and potential significance of atmospheric contamination by pesticides. The answers vary in their completeness, reflecting the strengths and weaknesses of existing information.

TABLE 1.1. Summary of review articles on physical and chemical atmospheric processes

Reference	Summary description
West, 1964	Summarized and presented examples of pesticide contamination (mostly DDT) of air, water, food, and humans. Noted that, at that time, there was only limited knowledge of the extent and significance of the environmental contamination by pesticides; there were no environmental or human monitoring systems, and the state of the science was technically unprepared to predict significant long-term effects of contamination by pesticides on humans and animals.
Middleton, 1965	Briefly discussed the presence of pesticides in the atmosphere, their sources, persistence, and the effects that application methods have on airborne pesticide concentrations. Showed the need for more study to evaluate the role and influence that pesticides have on air quality and the consequent effects of diminished air quality on man and the environment.
Crosby and Li, 1969	An in-depth review of the photochemical reactions and products of herbicide classes.
Finkelstein, 1969	Discussed the air pollution aspects of pesticides, but limited much of the discussion to health effects. Presented a limited discussion of pesticide sources and analytical methods.
Glotfelty and Caro, 1975	A brief discussion focused on agricultural sources, atmospheric transport and removal processes, and residence times. Assessed how the physical and chemical properties affect potential atmospheric distribution and transport potentials. Listed research needs.

TABLE 1.1. Summary of review articles on physical and chemical atmospheric processes-*Continued*

Reference	Summary description
Junge, 1975	Discussed the parameters that determine the atmospheric residence times of pesticides including transport and removal processes. Concluded that one of the major difficulties in understanding global transport and distribution of pesticides is the lack of knowledge of their atmospheric chemical behavior, which determines their residence times.
Crosby, 1976	An in-depth review of the photodecomposition process, methods and equipment needed for laboratory investigations, and the photochemistry of various herbicides and plant growth regulators.
Lewis and Lee, Jr., 1976	Discussed pesticide input sources into the atmosphere, occurrences and measured air concentrations in urban, rural, and indoor air, as well as transport and removal mechanisms.
Slinn, 1977	Presented semi-empirical formulae to estimate precipitation scavenging and dry deposition of particles and gases. Pesticides were not specifically addressed.
Glotfelty, 1978	Discussed the atmosphere as a sink for applied pesticides from agricultural settings, along with their atmospheric residence times and major atmospheric transformation reactions.
Bidleman and Christensen, 1979	A discussion focused on those factors that influence the atmospheric deposition processes of selected organochlorine pesticides (p,p'-DDT, chlordane, and toxaphene).
Sehmel, 1980	A review of particle and gas dry deposition processes. Summarized published measured and calculated deposition velocity values for a variety of inorganic particles. Discussed micrometeorological and surface variables that influence dry deposition removal rates. Pesticides were not specifically addressed.
Slinn and Slinn, 1980	Discussed a model to investigate the influence of particle growth by water vapor condensation on particle deposition.
Eisenreich and others, 1981	Discussed the wet and dry depositional processes for selected organochlorine pesticides. Presented concentration data for air and precipitation and calculated total deposition into the Great Lakes for these pesticides.
Atkinson and Carter, 1984	An in-depth review of gas-phase reaction rates of ozone with various classes of organic compounds and chemical functionalities under atmospheric conditions. Pesticides were not specifically addressed.
Murphy, 1984	Reviewed and discussed the atmospheric inputs of chlorinated hydrocarbons, including several pesticides, into the Great Lakes. Also discussed the importance of air/water exchange processes and mechanisms.
Barrie and Schemenauer, 1986	Discussed theoretical and observational approaches for understanding the mechanisms of pollutant wet deposition with respect to precipitation scavenging and fog deposition. Focused on wet deposition and acid-related substances.
Edwards, 1986	A general discussion of agrochemicals as environmental pollutants including reasons for pesticide use, their environmental effects, routes of human exposure and resulting effects, sociological and environmental factors, alternative pest control measures, and monitoring programs.

TABLE 1.1. Summary of review articles on physical and chemical atmospheric processes-*Continued*

Reference	Summary description
Finlayson-Pitts and Pitts, 1986	A comprehensive book on atmospheric chemistry including tropospheric photochemistry, experimental kinetics, reaction measurement techniques, and other aspects of transport and fate of organic pollutants. Pesticides were not specifically addressed at any length.
Pimentel and Levitan, 1986	Reviewed which crops the majority of pesticides were applied to, their application methods, and how they enter the water, soil, air, and biota as well as the effects on ecosystems. Estimated that less than 0.1 percent of all pesticides applied actually reaches the target pest. Also calculated a dollar cost versus benefit for pesticide use.
Seinfeld, 1986	A technical textbook providing a comprehensive review of the chemistry of air pollutants, the formation, growth, and dynamics of aerosols, the meteorology of air pollution, and the transport, diffusion, and removal of airborne pollutants. Pesticides were not specifically addressed.
Bidleman and Foreman, 1987	An in-depth review of the vapor-particle partitioning and distribution of semivolatile organic compounds in air. Investigated the characteristics of particulate matter in urban air and presented field and laboratory results.
Pankow, 1987	An in-depth, theoretical review of the partitioning behavior between vapor and aerosol particulate phases in the atmosphere.
Bidleman, 1988	Discussed how vapor-particle partitioning influences the atmospheric deposition processes, both wet and dry.
Bidleman and others, 1988	Reviewed the usage, and the atmospheric transport and deposition of toxaphene from the time of its high use to the present. Discussed air and precipitation concentrations reported in the literature along with analytical techniques. Discussed the physical and chemical properties of toxaphene and their relationship to the depositional processes.
Nicholson, 1988b	Reviewed a variety of experiments on resuspension of particles from various surfaces, and by mechanisms other than wind. Pesticides were not specifically addressed.
Schroeder and Lane, 1988	Discussed many of the important aspects in the overall fate of airborne pollutants including: Emission sources; atmospheric mixing and transport; photochemical transformations; and depositional processes. Pesticides were included as a subset of a larger group of organic pollutants.
Taylor and Glotfelty, 1988	A review of the factors that control the rate of volatilization of herbicides from soils and crops. Discussed basic physical processes and how physical placement affects volatilization. Presented several methods for estimating volatilization rates and results from several field studies.
Arimoto, 1989	A general review of the atmospheric deposition of chemical contaminants. Presented data on total inputs of selected organochlorine pesticides into each of the Great Lakes. Identified specific issues that require a better understanding for mass balance accounting of pollutants.
Atkinson, 1989	An in-depth review of gas-phase reaction rates of the hydroxyl radical with various classes of organic compounds and chemical functionalities under atmospheric conditions. Pesticides were not specifically addressed.

TABLE 1.1. Summary of review articles on physical and chemical atmospheric processes-*Continued*

Reference	Summary description
Davidson, 1989	A review of the current understanding of dry and wet deposition processes onto natural snow surfaces. Mathematical models were used to predict deposition rates and these predictions were compared to glacial record data. Pesticides were not specifically addressed.
Noll and Fang, 1989	Evaluated particle dry deposition fluxes, both toward and away from a surface, and airborne particle concentrations to estimate the effects of gravity and particle inertial deposition on atmospheric deposition velocities. Pesticides were not specifically addressed.
Cessna and Muir, 1991	Discussed types of photochemical reactions, photolytic studies of herbicides in water, air, and thin films. Presented photodegradation pathways and products for a variety of herbicide classes.
Tsai and others, 1991	Discussed the dynamic partitioning of semivolatile organic compounds in gas, particle, and rain phases during below-cloud rain scavenging. Pesticides were not specifically addressed.
Holsen and Noll, 1992	Compared actual field measured particle dry deposition using a variety of collection surfaces to model calculations using atmospheric particle size distribution data. Pesticides were not specifically addressed.
Iwata and others, 1993	Discussed the role of the ocean in understanding the long-range atmospheric transport and fate of organochlorine insecticides (DDTs, HCHs, and chlordanes) and PCBs. Estimated fluxes by gas exchange across the air/sea interface.
Pankow, 1994	An extended treatment on the theory of gas/particle partitioning of semivolatile organic compounds.
U.S. Environmental Protection Agency, 1994b	An informational report to congress which summarizes the current state of scientific knowledge on atmospheric deposition to the Great Waters (the Great Lakes, Lake Champlain, Chesapeake Bay, and coastal waters) of the United States. The report includes sections on effects, relative loadings, sources, recommendations, and actions.

CHAPTER 2

Characteristics of Studies Reviewed

All reviewed studies investigated pesticide occurrence in one or more atmospheric matrices (air, rain, snow, fog, aerosols). Table 2.1 summarizes selected characteristics of the studies reviewed. Each study is listed in chronological order of publication in Tables 2.2, 2.3, and 2.4 (at end of chapter) in one of three main categories: Process and matrix distribution studies (Table 2.2), state and local monitoring studies (Table 2.3), and national and multistate monitoring studies (Table 2.4). The sampling location(s) for each study is designated in corresponding Figures 2.1, 2.2, and 2.3 by the study number and an optional letter that differentiates the sampling locations if there is more than one for a study. Laboratory studies and review papers are cited in the text, as needed, but they are not included in Tables 2.2, 2.3, and 2.4.

Process and matrix distribution studies (Table 2.2, Figure 2.1) generally measured the concentration distributions of one or more pesticides between various atmospheric matrices to determine their physical and chemical properties, controlling processes, or in the development of sampling or analytical methodologies. Field studies that monitored one or more atmospheric dissipation processes of specific pesticides from specific applications are also included. Most studies involved relatively specialized sampling at one or several sites for several days, weeks, or months.

State and local pesticide monitoring studies (Table 2.3, Figure 2.2) were occurrence surveys for specific compounds or compound classes, usually at more than one site within a specific area, most typically within an area or region much smaller than the state in which they were done. This group includes a few studies with one location sampled over several months to several years, as well as studies with many locations sampled for several days, weeks, or months.

National and multistate pesticide monitoring studies (Table 2.4, Figure 2.3) were occurrence surveys for specific compounds or compound classes at more than one site in multiple states for several months to several years.

2.1 GENERAL DESIGN FEATURES

Several scales of study designs have been used to investigate pesticide occurrence in the atmosphere: local studies, which encompass areas of one to tens of square kilometers; regional studies, which encompass areas of tens to hundreds of square kilometers; and long-range studies, which encompass areas of hundreds to thousands of square kilometers. The local scale includes field studies that monitor pesticide drift during application, or the volatilization and off-site drift of applied compounds after application, or both. In these types of studies, the sampling frequency

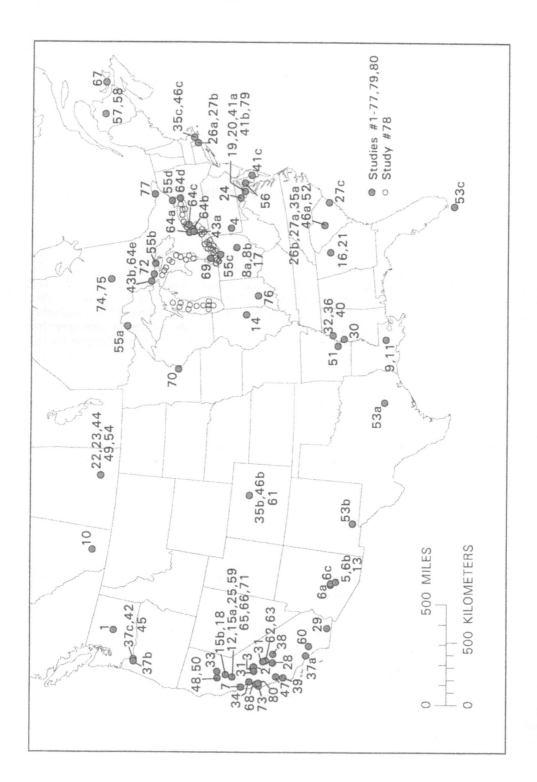

FIGURE 2.1. Sampling locations for pesticide process and matrix distribution studies listed in Table 2.2.

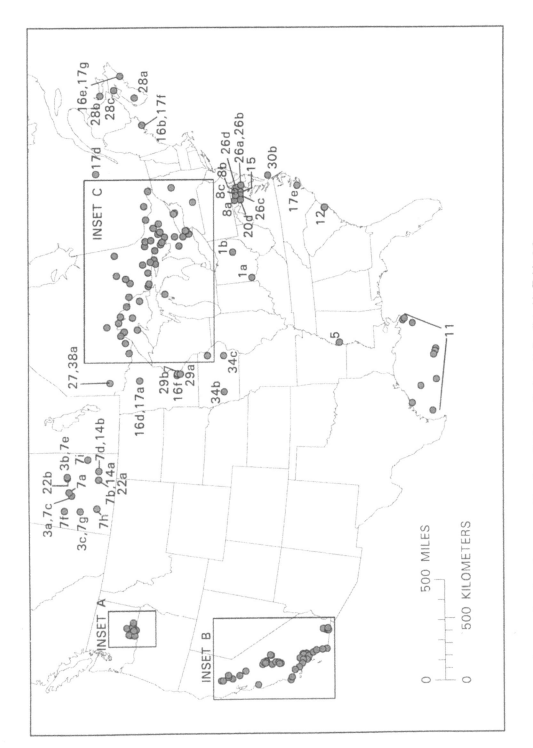

FIGURE 2.2. Sampling locations for state and local pesticide monitoring studies listed in Table 2.3.

FIGURE 2.2.—*Continued*

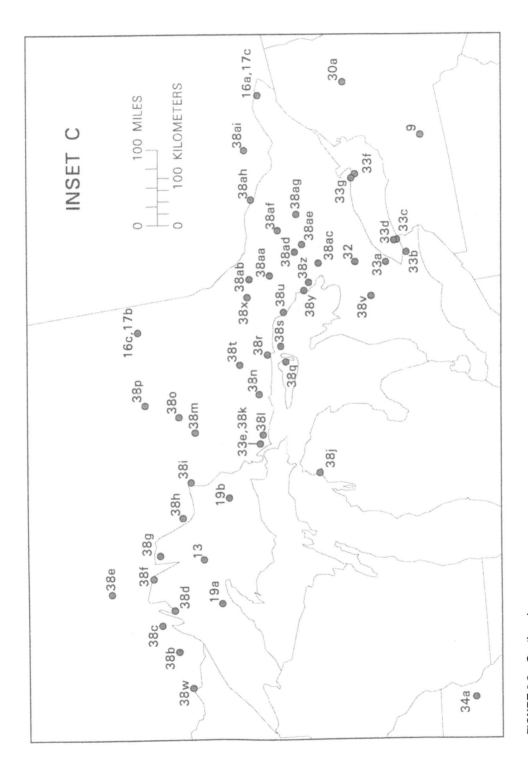

FIGURE 2.2.—*Continued*

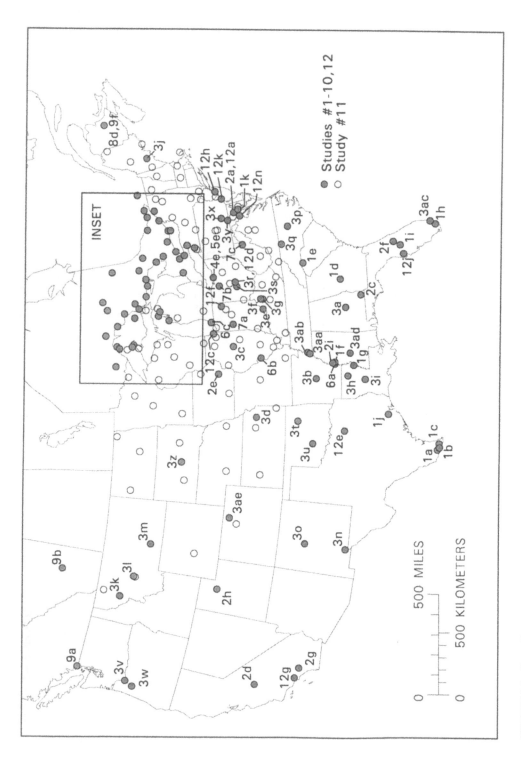

FIGURE 2.3. Sampling locations for national and multistate pesticide monitoring studies listed in Table 2.4.

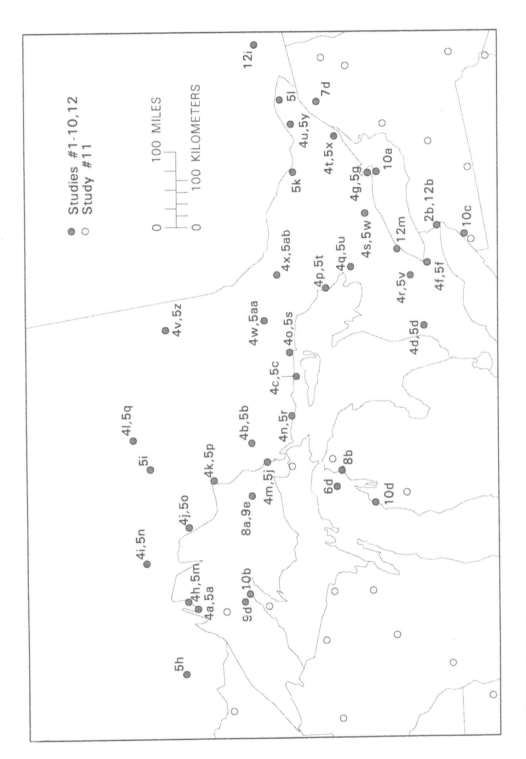

FIGURE 2.3.—*Continued*

Table 2.1. General study characteristics of pesticide studies

[<, less than]

Study characteristics	Study type		
	Process and matrix distribution studies	State and local monitoring studies	National and multistate monitoring studies
Number of studies	80	38	12
Number of sites			
Range	1-5	1-19	4-81
Median	1	3	6
Study duration (years)			
Range	0.02-2	<0.1-9	0.5-3
Median	<0.04	1	1
Sampled matrix (number of studies sampling)			
Air	63	25	7
Rain	8	16	8
Other	55	10	3
Compound class			
Organochlorine insecticides	37	19	10
Organophosphorus insecticides	12	12	5
Other insecticides	7	0	2
Herbicides	29	17	5

is usually very high. Samples are usually taken at intervals of 0.5 to 4 hours or more for several consecutive days of the study. The length of the sampling period depends on the information needed to fully assess the environmental behavior of the pesticide, the meteorological conditions, and the expected air concentrations. These types of studies usually last for 1 to 2 weeks and can generate hundreds of samples. The analytical methods usually are specific and optimized for one or several known compounds. Examples of these types of studies are listed in Table 2.2.

Local studies also include targeting the occurrence and distribution of locally used pesticides in one or more atmospheric matrices such as fog (Glotfelty and others, 1987; Turner and others, 1989; Schomburg and others, 1991; Seiber and others, 1993); air, (Arthur and others, 1976; Sava, 1985; Ross and Sava, 1986; Oudiz and Klein, 1988; Seiber and others, 1989; Ross and others, 1990; Fleck and others, 1991), and precipitation (Wu, 1981; Shulters and others, 1987; Glotfelty and others, 1990b; Capel, 1991). This type of sampling also can track the movement of one or more pesticides from high-use areas to low- or no-use areas (Muir and others, 1990; Nations and Hallberg, 1992; Zabik and Seiber, 1992). The range of pesticides analyzed for can be from one, such as the occurrence and distribution of parathion in three different use areas (Oudiz and Klein, 1988) to several pesticides used in or near the sampling areas (Seiber and others, 1989; Zabik and Seiber, 1992; Seiber and others, 1993) to multiclass/ multiresidue screenings (Glotfelty and others, 1987; Nations and Hallberg, 1992).

Sampling locations for regional studies are throughout a state or a large region. These types of studies can target analysis for one compound such as 2,4-D (Grover and others, 1976), toxaphene (Rice and others, 1986), or triallate (Grover and others, 1981); specific types of compounds such as the herbicides atrazine, simazine, alachlor, and metolachlor, used in corn production (Glotfelty and others, 1990b); or multiresidue/multiclass screening for a wide variety of compounds (Richards and others, 1987; Nations and Hallberg, 1992). These types of studies can last for 1 or more years and can generate hundreds of samples.

In the local- and regional-area type studies, knowledge of the individual pesticides and their use patterns for the area around each site is important in designing a sampling and analytical strategy to obtain the most complete picture of why the compounds detected in the atmosphere are there. Both study types can be designed to elucidate the spatial and temporal trends of one or many different pesticides. Large-scale, spatial trend studies such as those by Stanley and others (1971), Kutz and others (1976), Rice and others (1986), Richards and others (1987), and Goolsby and others (1994) require a different sampling strategy than local or field-scale studies. Reliable and identical sampling methods must be used at every location. This type of sampling network usually does not provide much detailed information on the long-range transport of pesticide as the samples are often weekly composites. They do provide a coarse indication of the distribution of pesticide occurrence around the area being sampled as each site is influenced by the local pesticide use.

Multiresidue screening for unknown compounds by chemical class usually requires a very large sample size and very low analytical detection limits. These types of studies are often used to monitor the background concentrations at trace levels (Giam and others, 1980; Seiber and others, 1989; Foreman and Bidleman, 1990; Knap and Binkley, 1991; Patton and others, 1991).

Studies in all categories were generally short-term, seldom lasting more than 1 year. Study designs ranged from monitoring airborne concentrations of a single pesticide near its application site to nationwide studies investigating concentrations in air and precipitation for a wide variety of pesticides. Generally, there was no consistency in sampling methodologies, sampling site placement, and collection timing and duration among studies. There also was no consistency in the selected analytes, analytical methods, or detection limits. Frequently, only compounds that were detected were reported. Those compounds that were analyzed for and not detected were reported in very few studies. Most studies were of the process and matrix distribution type, primarily due to the quantity of drift and post-application volatilization measurement studies. Most of the available data, however, for assessing the occurrence and distribution of pesticides in the atmosphere are from studies classified as state and local monitoring studies.

2.2 GEOGRAPHIC DISTRIBUTION

Figures 2.1, 2.2, and 2.3 show that the geographic distribution of sampling locations for the studies reviewed is highly uneven, with many areas of the nation never sampled, and others intensively sampled. The most extensive data collection efforts have been in the Northeast, the central Atlantic coastal areas, the Great Lakes, the Midwest, California, and Saskatchewan, Canada. Studies were also done in Mississippi, Washington, Hawaii, and the Gulf of Mexico. Most studies, however, have focused on sites in or near agricultural areas, resulting in a general bias toward this land use in understanding the atmospheric distribution of pesticides on a national scale.

2.3 MATRICES

Air has been the most sampled atmospheric matrix, particularly during the 1960's and 1970's. This may have been because sampling air to determine the occurrence of pesticides and their distribution between vapor and suspended particulate phases does not have the drawback of waiting for a specific event as is required for sampling rain, snow, or fog. During the 1980's, there was still much interest in air, but attention to precipitation and fog grew. New sampling and analytical methods have been developed that enable determination of pesticide concentration distributions between vapor and particles (Billings and Bidleman, 1983; Chang and others, 1985; Coutant and others, 1988, 1989; Lane and others, 1988; Johnson and others, 1990; Krieger and Hites, 1992; Turpin and others, 1993), between vapor and precipitation (Pankow and others, 1984; Chan and Perkins, 1989), or between vapor and fog (Glotfelty and others, 1987). Samples that are representative of the actual environment, however, are difficult to obtain, and much thought and work has gone into solving this problem (Keller and Bidleman, 1984; Van Vaeck and others, 1984; Coutant and others, 1988; Lane and others, 1988; Pankow, 1988; Ligocki and Pankow, 1989; Pankow and Bidleman, 1991; Zhang and McMurry, 1991; Cotham and Bidleman, 1992; Pankow and others, 1993; Turpin and others, 1993; Goss, 1993; Hart and Pankow, 1994).

2.4 TARGET ANALYTES

Most of pesticides investigated in the studies listed in Tables 2.2, 2.3, and 2.4 can be classified into four major groups: organochlorine insecticides, organophosphorus insecticides, triazine and acetanilide herbicides, and other herbicides. Published studies on other insecticides and fungicides were rare. The distribution of total sampling effort to each of these four groups is shown in relation to sampled matrix in Figure 2.4. In compiling the data for this figure, one study year was assigned for every year that the study took place regardless of starting month, number of sampling sites, sampling intensity, or duration for each group of compounds analyzed for in each matrix. For example, Nations and Hallberg (1992) analyzed rain at six sites in Iowa between October 1987 and September 1990 for a number of pesticides including six triazine and acetanilide herbicides, nine organophosphorus insecticides, and four other herbicides. The resulting study year assignment for each respective category was three each for the year grouping of 1980-1989 and one each for 1990-1993.

A great deal of effort has been expended on studying organochlorine pesticides since the mid-1970's (Figure 2.4) even though many of these compounds have been banned or their use greatly restricted in the United States. During the 1970's, the organophosphorus and triazine classes were moderately studied in air and precipitation. The 1980's showed an increase in the number of studies for these two classes, but these were relatively few when compared to the number of studies focusing on the organochlorine class.

Organochlorine compounds were the primary focus of the studies done on and around the Great Lakes while atrazine and several other corn herbicides were the main focus in the Midwest and Northeast. A wider variety of pesticides, including organophosphorus insecticides and a variety of herbicides, were found in California. Of the two national-scale studies that sampled in 20 or more states, Goolsby and others (1994) only analyzed rain for those herbicides used in corn and soybean production, primarily the triazines and acetanilides. Kutz and others (1976) did multiresidue analyses that included various organochlorine and several organophosphorus insecticides, and several chlorophenoxy acid herbicides in air. The Canadian studies generally focused on the organochlorine pesticides in the Great Lakes region with the exception of several studies that monitored the occurrence of selected herbicides used in wheat production in Saskatchewan (Que Hee and others, 1975; Grover and others, 1976, 1981, 1988a). Muir and others (1990) also analyzed for various herbicides in Ontario.

The analyses of organophosphorus insecticides and herbicides in the state and local and national and multistate studies were distributed fairly evenly. The airborne drift potential of the chlorophenoxy acid herbicide 2,4-D and its related esters was extensively studied during the late 1960's (Grover and others, 1976) and early 1970's (Que Hee and others, 1975; Farwell and others, 1976; Reisinger and Robinson, 1976). Since then, study efforts have shifted to other types of herbicides as well as the organochlorine and organophosphorus compounds in air and rain.

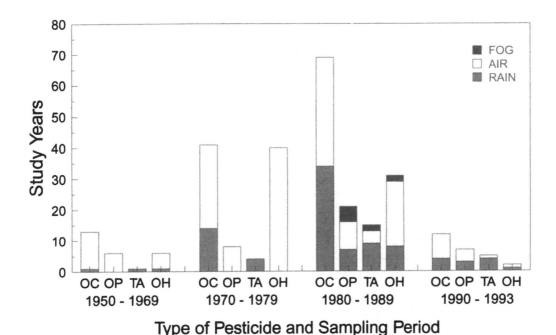

FIGURE 2.4. Sampling effort, in study years, per atmospheric matrix for the four major classes of pesticides from Tables 2.2, 2.3, and 2.4. Explanation: OC, organochlorine insecticides; OP, organophosphorus insecticides; TA, triazine and acetanilide herbicides; OH, other herbicides.

2.5 ANALYTICAL DETECTION LIMITS

A major problem that arises when comparing results of different studies is unknown or variable detection limits. Analytical detection limits were reported for less than half of the process and matrix distribution studies (Table 2.2). Several referred to companion manuscripts that detailed the sampling and analytical methodologies used, but even these references did not always contain the needed information. Detection limits for several compounds in several studies were inferred from the reported data when "<" (less than) values were given. In other cases, the lowest reported value for a compound or group of similar compounds was used as an estimate of the detection limit, but this did not always indicate the true study detection limits. Many detection limits in Tables 2.2, 2.3, and 2.4 were determined using the "<" values that were then applied to similar compounds within a class, where appropriate.

The variability in environmental sample sizes may be the reason specific detection limits are not reported in many of the studies in the literature. The analytical limits of detection for all pesticide classes in environmental samples are commonly determined by the sample size. If a lower detection limit is required, generally sample size must be increased, provided the sampling and extraction efficiencies remain the same.

Analytical limits of detection generally decreased by one to three orders of magnitude for all three classes of studies from the mid-1970's to the present. For those state and local monitoring studies that reported them, this decreasing trend was only evident for the organophosphorus insecticides detected in rain. The detection limits for organochlorine and herbicide pesticides reported in rain increased by two orders of magnitude during this same time period. The reasons for this are unclear.

Table 2.2. Characteristics and summaries of pesticide process and matrix distribution studies

[Letters under Occurrence refer to Location(s). Con, continued; Dep, deposition; ND, not detected; NR, not reported; OA, oxygen analog transformation of parent compound; PAHs, polycyclic aromatic hydrocarbons; PCBs, polychlorinated biphenyls; Ptcl, particulate matter; cm, centimeter; ft, feet; km, kilometer; <LOD, less than analytical limit of detection; m, meter; m^2, square meter; mi, mile; ng, nanogram; ng/m^3, nanogram per cubic meter; ng/smpl, nanogram per sample; pg, picogram; pg/m^3, picogram per cubic meter; μg, microgram; μg/kg, microgram per kilogram; μg/L, microgram per liter; μg/smpl, microgram per sample]

Study no.	Study	Sampling				Compounds			Comments
		Matrix	Date	Quantity	Location(s)	Name	Occurrence	Detection limit	
1	Batchelor and others, 1954	Air	NR	94	North Central Washington, near Wenatchee	Parathion	Field worker exposure experiment	NR	Air samples were taken as a part of a worker exposure study. Samples were taken at various locations including inside and outside of the orchard, at the mixing plant, warehouse, and mixing-loading areas as well as in nearby residential areas.
2	Caplan and others, 1956	Air	Summer, 1955	NR	Planada, CA	Malathion	Population exposure experiment	10 μg	Air samples were taken as part of an experiment to determine human exposures in populated areas during aircraft applications. Used a colorimetric analytical method and reported the sensitivity as 10 gamma; assumed this to be equivalent to 10 μg.

Table 2.2. Characteristics and summaries of pesticide process and matrix distribution studies--*Continued*

Study no.	Study	Sampling				Compounds			Comments
		Matrix	Date	Quantity	Location(s)	Name	Occurrence	Detection limit	
3	Culver and others, 1956	Air	Late spring, 1954	145 115	Merced, CA	Malathion Chlorthion	Field worker exposure experiment	10 μg	Air samples taken as a part of an experiment to determine human exposures at 3 downwind locations from an aerosol application to a field. Used same analytical method as Caplan and others, 1956.
4	Antommaria and others, 1965	Air	June-Dec 1964	10	Pittsburgh, PA	DDT, *p,p'*-	Population exposure experiment. Detected in 7 of 10 samples	NR	Sampled air for respirable and nonrespirable particulate matter that was analyzed for *p,p'*-DDT. DDD, DDE, or *o,p'*-DDT may have been present, but were not quantifiable.
5	Ware and others, 1969	Air Dep. Foli-age	Sept 1966	NR	Coolidge, AZ	Methoxychlor	Off-target drift measurement	NR	An experiment that compared the drift from simultaneous mist-blower and aircraft applications. The downwind air concentrations from the mist-blower were 6 times higher than the aerial application.
6	Frost and Ware, 1970	Air Dep. Foli-age	Aug 1965 May, Sept 1966 Aug 1967	NR	Chandler, AZ Coolidge, AZ Higley, AZ	Methoxychlor	Off-target drift measurement	NR	Four field experiments that compared drift results for high-clearance ground, mist-blower ground, and aerial application methods and weather conditions.

Table 2.2. Characteristics and summaries of pesticide process and matrix distribution studies--*Continued*

Study no.	Study	Sampling				Compounds			Comments
		Matrix	Date	Quantity	Location(s)	Name	Occurrence	Detection limit	
7	Murray and Vaughan, 1970	Air Dep.	Sept, Oct 1968	NR	San Francisco, CA	Chlorpyrifos Dieldrin Fluorescent particle	Off-target drift measurement	NR	The investigated material was applied to the ground in a single swath (single line-source) using a ground-based aerosol generator. Estimated pesticide deposition by counting fluorescent particle tracer deposition and the concentration of the drifting spray cloud. Measured drifting particles 21,000 ft downwind of a 0.5 mi line source.
8	Caro and others, 1971	Air	June-Oct 1968 (9 sampling days)	27	Coshocton, OH	Dieldrin	Field volatility experiment	NR	Pesticides incorporated 7.6 cm and seeded with corn. Lost 2.9% in 126 days by volatilization.
			May-Oct 1969 (7 sampling days)	51		Dieldrin Heptachlor		0.1 ng/m^3 0.1 ng/m^3	Incorporated 7.6 cm and seeded with corn. Lost 2.8 and 3.9%, respectively, in 167 days by volatilization.
9	Willis and others, 1971	Air Soil	Oct 1968 - Apr 1969 (13 sampling periods)	29	Baton Rouge, LA	DDT DDD	Field volatility experiment	NR	Incorporated to 15 cm and surface applied to three uncropped, flooded, nonflooded, and alternately flooded plots. Reported air concentrations and cumulative recoveries at 2 heights above each plot.

Table 2.2. Characteristics and summaries of pesticide process and matrix distribution studies--*Continued*

Study no.	Sampling					Compounds			Comments
	Study	Matrix	Date	Quan-tity	Location(s)	Name	Occurrence	Detection limit	
10	Grover and others, 1972	Air Dep.	NR	NR	Ralston, Alberta, Canada	2,4-D: Butyl ester Dimethylamine salt	Off-target drift measurement	2.9 milli-curie per millimole	Single line-source; ground rig application. Used ^{14}C labeled herbicide and measured liquid droplet drift for both esters between 3-4% of the mass sprayed. The butyl ester showed additional vapor drift of up to 33% of the mass applied within 30 minutes. Drift monitored to a downwind distance of 100 m.
11	Willis and others, 1972	Air Soil	Sept 1969-Jan 1970 (15 sampling periods)	45	Baton Rouge, LA	Dieldrin	Field volatility experiment	NR	Surface applied. Lost 18, 2 and 7% in 150 days by volatilization from continuously moist, flooded, and nonflooded plots, respectively.
12	Yates and others, 1974	Dep. Foli-age	Various times through-out 1966	NR	Davis, CA	Fluorescent particle Methoxychlor Tetradifon	Off-target drift measurement	NR	Single line-source; aircraft application. Correlated the downwind drift concentration with atmospheric stability and oil content of the spray formulation. Detected residues 0.5 mi downwind were 13 times greater during very stable atmospheric conditions than during neutral conditions.

Table 2.2. Characteristics and summaries of pesticide process and matrix distribution studies--*Continued*

Study no.	Study	Sampling				Compounds			Comments
		Matrix	Date	Quan-tity	Location(s)	Name	Occurrence	Detection limit	
13	Ware and others, 1974	Air Foli-age	July-Aug 1972	NR	Coolidge, AZ	Azodrin Methyl Parathion Parathion	Field worker exposure experiment	NR	A cotton field was sprayed with 3 insecticides and residues on foliage, skin, clothing, and air concentrations at breathing height were measured 24 hours after the application on 4 volunteers that stayed in the treated area for 5 hours.
14	Goering and Butler, 1975	Dep.	Fall, 1972 Spring, 1973	NR	Urbana, IL	Fluorescent particle	Off-target drift measurement	NR	Single line-source; ground rig application. Measured effectiveness of drift reducing formulation application technology. Used dual tracer technique to assess drift potential of each formulation under the same meteorological conditions.
15	Soderquist and others, 1975	Air Soil	June 1973 Apr 1974	14 10	Davis, CA A field in Sutter County, CA	Trifluralin Several photoproducts	Field volatility experiment	0.01 ppm	Experimentally determined several vapor-phase photoproducts of trifluralin in the laboratory and conducted 2 field experiments to determine if the photoproducts could be detected after surface and incorporated applications.

Table 2.2. Characteristics and summaries of pesticide process and matrix distribution studies--*Continued*

Study no.	Study	Sampling				Compounds			Comments
		Matrix	Date	Quan-tity	Location(s)	Name	Occurrence	Detection limit	
16	Harper and others, 1976	Air	June-Oct 1973 (9 sampling days) and June-July 1974	38	Watkinsville, GA	Trifluralin HCH, γ- (1974)	Field volatility experiment	5-6 ng/m^3 (4 hour sampling periods) 2-3 ng/m^3 (12 hour sampling periods)	Incorporated to 2.5 cm and seeded with soybeans. Reported microclimate effects on volatilization from crop. Volatilization losses reported in White and others, 1977.
17	Taylor and others, 1976	Air Soil	April-Oct 1969 (8 sampling days)	52	Coshocton, OH	Dieldrin Heptachlor	Field volatility experiment	0.1 ng/m^3 0.1 ng/m^3	Incorporated to 7.5 cm and seeded with corn. Lost 4 and 7%, respectively, in 170 days by volatilization.
18	Soderquist and others, 1977	Air Soil Water	May-June 1975	NR	Sutter County, CA	Molinate	Field volatility experiment	Reported as: "ng/m^3 range"	Surface applied to flooded rice field. Lost 78% in 7 days by volatilization.
19	Taylor and others, 1977	Air Soil Crop	July-Aug 1973 (7 sampling days)	NR	Beltsville, MD	Dieldrin Heptachlor	Field volatility experiment	0.1 ng/m^3 0.1 ng/m^3	Surface applied to short grass. Lost 12 and 46%, respectively, in 0.5 days and 23 and 54% in 3 days by volatilization.
20	Turner and others, 1977	Air Soil Crop	July-Oct 1973 (9 sampling days)	NR	Beltsville, MD	Dieldrin Photodieldrin	Field volatility experiment	NR	Surface applied to short grass, same experiment as Taylor and others, 1977. Reported photodieldrin volatilization rate to be 1/5 that of dieldrin.

Table 2.2. Characteristics and summaries of pesticide process and matrix distribution studies--*Continued*

Study no.	Sampling					Compounds			Comments
	Study	Matrix	Date	Quantity	Location(s)	Name	Occurrence	Detection limit	
21	White and others, 1977	Air Soil	June-Oct 1973 (9 sampling days)	38	Watkinsville, GA	Trifluralin	Field volatility experiment	5-6 ng/m^3 (4 hour sampling periods) 2-3 ng/m^3 (12 hour sampling periods)	Incorporated to 2.5 cm and seeded with soybeans. Lost 20 and 22% in 35 and 120 days, respectively, by volatilization.
22	Grover and others, 1978	Air Dep.	June 1974	NR	Regina, Saskatchewan, Canada	2,4-D amine Fluorescent particle	Off-target drift measurement	1 ppm	Single line-source; ground rig application. Conducted a mass balance assessment of drift from a herbicide application.
23	Maybank and others, 1978	Air Dep.	No dates given	NR	Regina, Saskatchewan, Canada	2,4-D: Butyl ester Octyl ester	Off-target drift measurement	NR	Single line-source; ground rig and aircraft applications. Compared off-target drift from both application methods, both during and post application.
24	Turner and others, 1978	Air Soil	May-July 1976 (6 sampling days)	26	Frederick, MD	Chlorpropham	Field volatility experiment	5 ng/m^3	Surface applications of emulsifiable concentrate (EC) and microencapsulated formulations to different fields seeded with soybeans. Estimated nearly 50 and 25% of the soil residue loss, respectively, was due to volatilization.

Table 2.2. Characteristics and summaries of pesticide process and matrix distribution studies—*Continued*

Study no.	Study	Sampling				Compounds			Comments
		Matrix	Date	Quantity	Location(s)	Name	Occurrence	Detection limit	
25	Yates and others, 1978	Air Dep. Crop	NR	NR	Davis, CA	Glyphosate	Off-target drift measurement	NR	Single line-source; fixed-wing aircraft and helicopter applications. Compared drift from three application methods using different spray nozzles and a formulation thickening agent.
26	Christensen and others, 1979	Dep.	Feb 1975-May 1975 Dec 1976-May 1977	5 9	a Kingston, RI b Columbia, SC	Chlordane, *cis-* Chlordane, *trans-* DDT, *p,p'-*	b b a,b	NR	Developed a sampling technique that mimicked as closely as possible the collection characteristics of a water surface. Compared the collection characteristics of several hydrophilic surfaces to those of a dry pan. Calculated deposition fluxes for several organochlorine compounds.
27	Bidleman and Christensen, 1979	Air Rain Dry-dep.	Sept 1973-May 1975 Dec 1976-April 1979 July 1977-April 1979	NR	a Columbia, SC b Kingston, RI c North Inlet, SC	Chlordane DDT Toxaphene	a, c a, b, c a, c	NR	Investigated ambient concentrations and the relationship between particle deposition velocity (Vd) and vapor-particle partitioning. Only air concentration data given for each sampling period along with washout ratios. Included dry deposition data from Christensen and others, 1979, and toxaphene wet deposition data from Harder and others, 1980.

Table 2.2. Characteristics and summaries of pesticide process and matrix distribution studies--*Continued*

Study no.	Study	Sampling				Compounds			Comments
		Matrix	Date	Quan-tity	Location(s)	Name	Occurrence	Detection limit	
28	Seiber and others, 1979	Air Soil Crop	Aug-Oct 1975 (5 sampling days)	NR	Corcoran, CA	Toxaphene	Field volatility experiment	NR	Surface applied to mature cotton plants. Concluded that volatilization was a major dissipative route from leaf surfaces and soil.
29	Cliath and others, 1980	Air Water	May 1977	26	Brawley, CA	EPTC	Field volatility experiment	NR	Surface applied in flood irrigation. Lost 74% in 2.2 days by volatilization.
30	Hollings-worth, 1980	Air	1976 (for 118 days) 1977 (for 103 days) (4 and 14 day sampling periods)	NR	Stoneville, MS	Trifluralin	Field volatility experiment	NR	Incorporated to 7.6 cm and seeded with soybeans. Loss of 0.32 and 0.45% in 120 days by volatilization for both years, respectively.
31	Oshima and others, 1980	Air Dep.	Sept, Oct 1979	NR	Fresno and Merced Counties, CA	DEF Folex	Off-target drift measurement	NR	Field-source; aerial application. Monitored off-site drift from daylight and night applications to commercial cotton fields. Also monitored ambient air at two nearby residential communities.
32	Willis and others, 1980	Air Soil Crop	Sept 1974	24	Clarksdale, MS	Toxaphene	Field volatility experiment	15 ng/m^3	Three surface applications to mature cotton field. Lost 25% in 5 days by volatilization after last application.

Table 2.2. Characteristics and summaries of pesticide process and matrix distribution studies--*Continued*

Study no.	Study	Sampling			Location(s)	Compounds			Comments
		Matrix	Date	Quan-tity		Name	Occurrence	Detection limit	
33	Crosby and others, 1981	Air Foli-age	June-Aug 1979	NR	Butte County, CA	MCPA dimethyl amine salt	Off-target drift measurement	3 ng/m^3	Field-source; aerial application. Monitored off-site drift from a commercial application to a rice field and drift residues on nearby orchard foliage. Spray drift was measured to 400 m downwind.
34	Oshima and others, 1982	Air Water Dep.	July-Aug 1981 Oct 1981	NR	Santa Clara Valley, CA a 250 mi^2 area	Malathion Malathion OA	Off-target drift measurement	NR	Area-source; aerial application. Characterized the impact of multiple aerial applications at 300 ft on an urban environment.
35	Billings and Bidleman, 1983	Air	1977-80 Jan 1980 June 1980	NR	a Columbia, SC b Denver, CO c New Bedford, MA	Chlordane DDE, *p,p'-* HCB HCH (α+γ) Toxaphene	a, b, c a, b a, b, c a, b, c a	NR	Comparative sampling using 3 trapping media. Investigated the trapping efficiency of each media with respect to air volume sampled and ambient temperature. Found that toxic air concentrations increased 10 times from Aug-Sept 1977 to 1979.
36	Harper and others, 1983	Air Soil Crop	Aug-Sept 1976 (11 sampling days)	109	Clarksdale, MS	Toxaphene DDT	Field volatility experiment	15 ng/m^3 1 ng/m^3	Explained volatilization losses from plant and soil surfaces in terms of micrometeorological conditions.

Table 2.2. Characteristics and summaries of pesticide process and matrix distribution studies--*Continued*

Study no.	Study	Sampling				Compounds			Comments
		Matrix	Date	Quan- tity	Location(s)	Name	Occurrence	Detection limit	
37	Pankow and others, 1983	Rain Air	Feb-Mar 1982 Oct 1982	NR	a Los Angeles, CA (rain) b Beaverton, OR (rain) c Portland, OR (air, rain)	HCH, α-HCH, γ-	a,b,c a	NR	Sampled and analyzed rain in several different ways to attempt to distinguish between the dissolved and scavenged particle-associated levels of a variety of organic compounds, mostly PAHs.
38	Segawa and others, 1983	Air Water Dep.	Nov 1982	NR	Tulare County, CA (a 14 acre area)	Carbaryl	Off-target drift measurement	NR	Field-source; aerial application. Determined the efficacy of a potential gypsy moth pesticide application over an urban environment. Measured fallout and drift from applications at 120 and 250 ft elevation. Measured drift at 80 and 550 m downwind of application site.
39	Weaver and others, 1983	Air Dep.	Mar 1982	NR	San Luis Obispo, CA (15 miles south of city)	Carbaryl	Off-target drift measurement	NR	Line-source; aerial application. Measured fallout and drift from applications at 50 and 100 ft elevations. Measured drift to 350 ft upwind and downwind of application swath.

Table 2.2. Characteristics and summaries of pesticide process and matrix distribution studies--*Continued*

Study no.		Sampling				Compounds			Comments
	Study	Matrix	Date	Quan-tity	Location(s)	Name	Occurrence	Detection limit	
40	Willis and others, 1983	Air Soil Crop	Aug-Sept 1976 (11 sampling days)	109	Clarksdale, MS	Toxaphene DDT	Field volatility experiment	15 ng/m^3 1 ng/m^3	Two surface applications to 50 cm high cotton plants at 4 day intervals. Lost 17% of the amount intercepted by crop from first toxaphene application in 10.7 days by volatilization. Lost 37 and 36% of the plant intercepted toxaphene and DDT from the second application in 10.7 days by volatilization. Same experiment as Harper and others, 1983.
41	Glotfelty and others, 1984	Air Soil	Aug 1975 June 1977 June 1978	24 18 18	a Beltsville, MD b Beltsville, MD c Salisbury, MD	Chlordane Dacthal Heptachlor HCH, γ- Trifluralin	a,c a a,c b,c a,b,c (field volatility experiment)	For all compounds: <1 ng/m^3	Surface applied to moist or dry soil. Volatilization losses: a: Lost 50% in 2.5, 1.4, 0.25, and 0.31 days, respectively. b: Lost 50 % in 0.25 and 0.13 days and 90% in 6 and 2.5 days, respectively. c: Lost 2, 14, 12, and 2 % in 2.1 days, respectively.
42	Pankow and others, 1984	Rain	Mar-April 1982 Oct-Dec 1982	NR	Portland, OR	HCH, α- HCH, β- HCH, γ- HCH, δ- HCB	Rain <LOD Rain <LOD <LOD	0.50 ng/L 0.50 ng/L 0.50 ng/L 0.50 ng/L 0.10 µg/L	Investigated various compounds in air and rain and the relation to Henry's law values and temperature.

Table 2.2. Characteristics and summaries of pesticide process and matrix distribution studies—*Continued*

Study no.	Sampling				Location(s)	Compounds			Comments
	Study	Matrix	Date	Quantity		Name	Occurrence	Detection limit	
43	Strachan and Huneault, 1984	Rain	April-May and July-Aug 1981	NR	a 50 Mile Point, Ontario, Canada b Turkey Lakes, Ontario, Canada	Chlordane, *cis-*	ND	1.0 ng/smpl	Described sampler designed to collect and concentrate persistent organic substances in rain in either particulate or dissolved forms. Averaged recoveries for a variety of organochlorine insecticides was 86%. Placed triplicate samplers at each location and tested recoveries for XAD-2 and XAD-7 resins.
						Chlordane, *trans-*	a, b	1.0 ng/smpl	
						DDD, *p,p'-*	ND	4.0 ng/smpl	
						DDE, *p,p'-*	ND	1.0 ng/smpl	
						DDT, *p,p'-*	b	3.0 ng/smpl	
						Dieldrin	b	1.0 ng/smpl	
						Endosulfan	a, b	1.5 ng/smpl	
						Endrin	b	2.0 ng/smpl	
						HCB	ND	0.5 ng/smpl	
						HCH, α-	a, b	0.5 ng/smpl	
						HCH, γ-	a, b	0.5 ng/smpl	
						Heptachlor epoxide	ND	1.0 ng/smpl	
						Methoxychlor	ND	5.0 ng/smpl	
						Mirex	ND	2.0 ng/smpl	
44	Grover and others, 1985	Air Soil Crop	June 1979	79	Regina, Saskatchewan, Canada	2,4-D, iso octyl ester	Field volatility experiment	NR	Surface applied to 20 cm high spring wheat. Lost 21% in 5 days by volatilization. Measured application drift losses of <0.2%.
45	Ligocki and others, 1985a	Air Rain	Feb-April 1984	7	Portland, OR	DDE, *p,p'-*	Air	NR	Analyzed for various organic compounds in local air and rain and related the findings to Henry's law values and temperature.
						HCB	Air		
						HCH, α-	Air, Rain		
						HCH, γ-	Rain		

Table 2.2. Characteristics and summaries of pesticide process and matrix distribution studies--*Continued*

Study no.	Study	Sampling				Compounds			Comments
		Matrix	Date	Quan-tity	Location(s)	Name	Occurrence	Detection limit	
46	Bidleman and others, 1986	Air	1977-82 Jan, June 1980	NR	a Columbia, SC b Denver, CO c New Bedford, MA	Chlordane DDE, p,p'- DDT, p,p'- HCB HCH, α-	a, b a, b a, b a, b a, b	NR	Estimated the vapor-particle partitioning of semivolatile organochlorine compounds from field samples. Same data as study 26, but includes newer data from 1981-82.
47	Ross and others, 1986	Air	Feb 1985	NR	Paso Robles, CA	MCPA	Off-target drift measurement	0.27 µg/high-volume sample 0.42 µg/low-volume sample	Field-source; aerial application. Monitored off-site drift from a commercial application to a barley field. Study compared air samples taken at 5 minute intervals at two different sampling rates to longer sampling intervals.
48	Seiber and others, 1986	Air Soil Water	May 1985	NR	Willows, CA	Molinate	Field volatility experiment	NR	Surface applied as granules to flooded rice field. Lost 34% in 4 days by volatilization.
49	Smith and others, 1986	Air	June 1982	46	Regina, Saskatchewan, Canada	Diclofop-methyl	Field volatility experiment	0.05 µg/m^3	Surface applied to 23 cm high wheat seedlings. Negligible volatility losses.
50	Ross and Sava, 1986	Air Water Soil Crop	May 1983	NR	Willows, CA	Molinate Thiobencarb	Field volatility experiment	NR	Surface applied as granules to seeded, flooded rice field. Lost 9 and <1%, respectively, in 6 days by volatilization.

Table 2.2. Characteristics and summaries of pesticide process and matrix distribution studies--*Continued*

Study no.	Study	Matrix	Date	Quantity	Location(s)	Name	Occurrence	Detection limit	Comments
								(Sampling)	
						(Compounds)			
51	Barnes and others, 1987	Air Dep. Crop	May 1985	NR	Dumas, AR	Propanil	Off-target drift measurement	0.003 μg/mL	Field-source; aerial application. Monitored off-site drift during and after application to determine the potential deposition and inhalation concentrations to human workers. Also measured drift residues and phytotoxic effects to adjacent crops. Reported detection limits as per final sample extract volume.
52	Foreman and Bidleman, 1987	Air	Oct-Nov 1984 Apr, Aug 1985	NR	Columbia, SC	Chlordane, cis-Chlordane, trans-DDE, p,p'-DDT, p,p'-HCB HCH, α-HCH, γ-Nonachlor, trans-Toxaphene	Not quantifiably collected	NR	Laboratory and field experimental results for an experimental system for investigating the vapor-particle partitioning of trace organic pollutants.
53	Atlas and Giam, 1988	Air Rain	Feb 1979-Feb 1980 1982	NR	a College Station, TX (air, rain) b White Sands, NM (air only) c Pigeon Key, FL (rain only)	Chlordane DDE, p,p'-DDT, p,p'-Dieldrin HCB HCH, α-HCH, γ-Toxaphene	a, b a, b, c a, b, c a a, b a, b a, b a, b	0.01-0.05 ng/L (inferred from data)	Investigated the ambient concentrations of organo-chlorine compounds and their relationship to vapor/particle partitioning and vapor/particle scavenging by rain.

Table 2.2 Characteristics and summaries of pesticide process and matrix distribution studies--*Continued*

Study no.	Sampling					Compounds			Comments
	Study	Matrix	Date	Quantity	Location(s)	Name	Occurrence	Detection limit	
54	Grover and others, 1988b	Air Soil	May-Oct 1983 (71 sampling days)	370	Regina, Saskatchewan, Canada	Triallate Trifluralin	Field volatility experiment	NR	Incorporated to 5 cm in a wheat seeded field. Lost 9.8 and 12.2 % in 7 days and 17.6 and 23.7% in 67 days, respectively. Measured application drift losses of <1%.
55	Chan and Perkins, 1989	Rain Snow	1986	93	Sites in Ontario, Canada: a Sibley Province Park b South Baymouth c Pelee Island d Wolfe Island	Aldrin	a, c, d	NR	Tested precipitation sampler design in the Great Lakes Basin.
						Chlordane, *cis-*	All		
						Chlordane, *trans-*	All		
						DDD, p,p'-	All		
						DDE, p,p'-	All		
						DDT, o,p'-	a, c		
						DDT, p,p'-	b, c, d		
						Dieldrin	All		
						Endosulfan, I	All		
						Endosulfan, II	All		
						Endrin	c, d		
						HCH, α-	All		
						HCH, γ-	All		
						Heptachlor	c		
						Heptachlor epoxide	a, c, d		
						Methoxychlor	All		
						Mirex	ND		
56	Glotfelty and others, 1989	Air	May-June 1981 (8 sampling days)	39	Near the Wye River, MD	Alachlor Atrazine Simazine Toxaphene	Field volatility experiment	25 ng/m^3 5 ng/m^3 5 ng/m^3 NR	Surface applied to seeded corn field. Lost 19, 2.4, 1.25, and 31% in 21 days, respectively, by volatilization.

Table 2.2. Characteristics and summaries of pesticide process and matrix distribution studies--*Continued*

Study no.	Sampling				Compounds			Comments
	Study	Matrix	Date	Quan-tity	Location(s)	Name	Occurrence	Detection limit	

Study no.	Study	Matrix	Date	Quan-tity	Location(s)	Name	Occurrence	Detection limit	Comments
57	Riley and Wiesmer, 1989	Dep.	July-Aug 1986	NR	New Brunswick, Canada	Deltamethrin	Off-target drift measurement	NR	Field-source; aerial application. Monitored off-site drift from a commercial application to potato fields. Determined the extent of necessary off-site drift buffer zone. Measured downwind drift to 100 m.
58	Riley and others, 1989	Air	NR	80	New Brunswick, Canada	Aminocarb	Off-target drift measurement	<0.1 μg/m^3	Measured mean air concentrations 2-3 km downwind of an aerial application to a forest canopy. Compared results of applications made during different meteorological conditions.
59	Ross and others, 1989	Air Dep. Crop	Apr-Aug 1987	NR	Davis, CA	Dacthal	Off-target drift measurement	0.5 ng/m^3 (air) 20 μg/kg (soil and vegetation)	Field-source; ground rig application. Study designed to determine the contributions of application drift, volatilization, and wind erosion to off-site contamination.

Table 2.2. Characteristics and summaries of pesticide process and matrix distribution studies--*Continued*

Study no.	Study	Sampling					Compounds			Comments
		Matrix	Date	Quantity	Location(s)		Name	Occurrence	Detection limit	
60	Clendening and others, 1990	Air Soil	NR	NR	A 2-hectare field in southern California		Atrazine Bromacil EPTC Prometon Triallate	Field mass-balance study	NR	Surface application and monitoring of 5 herbicides for 1 month to evaluate their fate under identical environmental conditions and to compare measured results of volatilization, leaching, and degradation to screening model predictions.
61	Foreman and Bidleman, 1990	Air	Oct 1985 Jan 1986	4 5	Denver, CO		DDE, *p,p'-* DDT, *p,p'-* Chlordane, *trans-* Chlordane, *cis-* Nonachlor, *trans-*	both dates both dates both dates both dates both dates	NR (0.001 ng/m^3; inferred from data)	Investigated the distribution of semivolatile organic compounds including several organochlorine pesticides between vapor and particulates. Related this partitioning to the ambient temperature and vapor pressure.
62	Glotfelty and others, 1990a	Air Fog	Jan 8-13, 1986	6	Parlier, CA		Chlorpyrifos Chlorpyrifos OA Diazinon Diazinon OA Methidathion Methidathion OA Parathion Parathion OA	All events sampled. (Chlorpyrifos OA was not found in Jan 11 sample.)	NR	Selectively looked for four organophosphorus compounds and their oxygen analogs (oxons) to determine their fog-air distribution. Presented concentrations in fog water only and percent distribution between air, water and particles for Jan 12 only.

Table 2.2. Characteristics and summaries of pesticide process and matrix distribution studies--*Continued*

Study no.	Study	Sampling Matrix	Sampling Date	Sampling Quantity	Location(s)	Compounds Name	Compounds Occurrence	Detection limit	Comments
63	Glotfelty and others, 1990c	Air Soil Tree-rinse	Jan 13-17, 1989	7	Parlier, CA	Diazinon	Field volatility experiment	NR	A field experiment that determined the spray distribution, drift, and volatilization from an air-blast spray application to a dormant peach orchard.
64	Johnson and others, 1990	Air	1985-87	NR	a Sheridan Park, Mississauga, Ontario, Canada b Stoney Creek, Ontario, Canada c Lake Ontario near Niagara d Lake Ontario near Sand Banks e Turkey Lakes, Ontario, Canada	HCB HCH, α- HCH, γ-	All All All	13 pg/m^3 3 pg/m^3 12 pg/m^3	Investigated three compounds occurrence in air while testing a gas and particulate sampler. Also looked at their phase distribution. Found that HCB concentration is constant throughout the year while α-HCH concentration varies directly with temperature and γ-HCH concentrations were highest in the spring and decreased with temperature.
65	Majewski and others, 1990	Air Soil	Apr 1985 (6 sampling days)	13	Davis, CA	Chlorpyrifos Diazinon HCH, γ- Nitrapyrin	Field volatility experiment	1.0 ng/m^3 0.3 ng/m^3 10 ng/m^3 10 ng/m^3	Surface applied. Lost 0.2, 0.1, 6.6, and 5.5%, respectively, in 4 days by volatilization.
66	Ross and others, 1990	Soil Crop	Sept 1987 (9 sampling days)	19	Davis, CA	Dacthal	Field volatility experiment	NR	Surface applied as wettable powder. Lost 10% in 21 days by volatilization based on 336 days regression of soil loss.

Table 2.2. Characteristics and summaries of pesticide process and matrix distribution studies--*Continued*

| Study no. | Study | Sampling | | | | Compounds | | | Comments |
		Matrix	Date	Quan-tity	Location(s)	Name	Occurrence	Detection limit	
67	Ernst and others, 1991	Dep.	NR	NR	Albany Corner, Prince Edward Island, Canada	Endosulfan	Off-target drift measurement	NR	Field-source; aerial application. Monitored off-site drift from a commercial application to potato fields. Containers of water were used as deposition samplers and the water was subsequently used in laboratory toxicity testing on water biology. Measured downwind drift to 200 m.
68	Fleck and others, 1991	Dep.	Nov 1988	NR	Moss Landing, CA	Endosulfan	Off-target drift measurement	10 µg/smpl	Field-source; aerial application. Monitored off-site drift from a commercial application to artichoke fields. Measured drift at 5.5 m from edge of field and found only 0.5-2.0% of measured application rate.
69	Frank and others, 1991	Dep. Soil Water Sedi-ment Fish	Aug 1984	NR	Tecumseh, Ontario, Canada	Permethrin	Off-target drift measurement	10 µg/kg (fish) 5 µg/kg (soil) 5 µg/kg (sediment) 0.002 µg/L (water) 0.1 µg/filter disk	Field-source; aerial application. Monitored off-site drift from a commercial application to potato fields. Study was designed to determine effective and practical buffer zones for protection of sensitive and productive bodies of water.

Table 2.2. Characteristics and summaries of pesticide process and matrix distribution studies--*Continued*

Study no.	Sampling				Compounds			Comments
	Study	Matrix	Date	Quan-tity	Location(s)	Name	Occurrence	Detection limit	
70	Franz and others, 1991	Rain	1986	NR	50 km north of Minneapolis, MN	9 Chlorinated pesticides	Only reported p,p'-DDT	NR	Evaluated four precipitation samplers collection efficiencies. Reported data for p,p'-DDT only. Most of the effort was focused on PAHs and PCBs.
71	Majewski and others, 1991	Air	Sept 1987 (9 sampling days)	19	Davis, CA	Dacthal	Field volatility experiment	NR (<0.001 $\mu g/m^3$ by inference)	Surface applied as wettable powder. Lost 36% in 21 days by volatilization based on 84 days regression of soil loss. Same experiment as Ross and others, 1990.
72	Payne and others, 1991	Dep. Foliage	NR	NR	Thessalon, Ontario, Canada	Permethrin	Off-target drift measurement	0.05 $\mu g/L$	Field-source; aerial application. Monitored downwind air and deposition concentrations to 250 m from an application to a forest canopy. Deposition concentrations were used to estimate biological effects to water bodies. Spray buffer zones were also determined.

Table 2.2. Characteristics and summaries of pesticide process and matrix distribution studies--*Continued*

Study no.	Study	Sampling				Compounds			Comments
		Matrix	Date	Quantity	Location(s)	Name	Occurrence	Detection limit	
73	Schomburg and others, 1991	Air Fog	Sept 1987	6 each matrix	3 sites along the Pacific coast near Monterey, CA	Carbaryl	All	NR	Compared pesticide concentrations in fog in agricultural and non-agricultural areas along the coast. Also compared the pesticide content and concentrations of the coastal fog to the Central Valley fog. Presented concentrations in fog water only and percent distribution between air and water phases for Sept 20 only.
						Chlorpyrifos	All	0.014 µg/L	
						Chlorpyrifos OA	All	NR	
						Diazinon	All	NR	
						Diazinon OA	All	NR	
						Fonofos	All	0.012 µg/L	
						Fonofos OA	All	NR	
						Malathion	All	NR	
						Malathion OA	All	NR	
						Methidathion	All	0.040 µg/L	
						Methidathion OA	All	0.012 µg/L	
						Methyl Parathion	All	0.049 µg/L	
						Methyl Parathion OA	All	NR	
						Parathion	ND	0.010 µg/L	
						Parathion OA	All	NR (inferred from data)	
74	Payne, 1992	Dep.	July, Aug 1987	NR	Lowther, Ontario, Canada	Glyphosate	Off-target drift measurement	500 ng	Field-source; aerial application. Monitored off-site drift from a commercial application to a clear-cut forest. Compared off-site deposition concentrations from 3 different application dispersal systems. Measured downwind drift to 300 m.

Table 2.2. Characteristics and summaries of pesticide process and matrix distribution studies--*Continued*

Study no.	Study	Sampling					Compounds			Comments
		Matrix	Date	Quantity	Location(s)	Name	Occurrence	Detection limit		
75	Payne and Thompson, 1992	Dep.	July, Aug 1987	NR	Lowther, Ontario, Canada	Glyphosate	Off-target drift measurement	500 ng		Field-source; aerial application. Monitored off-site drift from a commercial application to a clear-cut forest. Same field experiment as Payne (1992), but focused on the drift from a single dispersal system and the drift effects under various meteorological conditions. Measured drift to 300 m.
76	Burgoyne and Hites, 1993	Air	Jan 1991-Feb 1992 (3 times per month for 14 months)	NR	Bloomington, IN	Endosulfan I Endosulfan II	All July 18 and Aug 7	0.3 pg (average sample = 1,140 m^3)		Investigated the occurrence and air concentrations of endosulfan and the relation with temperature and wind direction.
77	Majewski and others, 1993	Air Soil	Sept 1990 (6 sampling days)	40	Ottawa, Ontario, Canada	Triallate Trifluralin	Field volatility experiment	NR <0.01 µg/m^3 (inferred from data)		Surface applied. Lost 74 and 54%, respectively, in 5 days by volatilization.
78	McConnell and others, 1993	Air	June 1989, Aug 1990	NR	a Green Bay, WI b Lake Michigan c Lake Huron d Lake Erie e Lake Ontario	HCH, γ- HCH, α-	All All	2.31 and 0.48 ng 0.78 and 0.03 ng		Sampled water and air for α- and γ-HCH to determine the gas exchange rate for HCHs in the Great Lakes. Sampled air using polyurethane foam (PUF) and PUF/Tenax. The two different air sampling media had different trapping efficiencies.

Table 2.2. Characteristics and summaries of pesticide process and matrix distribution studies--*Continued*

Study no.	Sampling				Compounds			Comments
	Study	Matrix	Date	Quan-tity	Location(s)	Name	Occurrence	Detection limit	
79	Whang and others, 1993	Air Soil Crop	April-May 1990 (10 sampling days)	NR	Beltsville, MD	Atrazine Chlorpyrifos Fonophos	Field volatility experiment	For all compounds: 0.01 $\mu g/m^3$	Surface applied to fallow tilled and untilled soil. Lost 0.9, 23, and 48%, respectively, in 4 days from the no-till plot, and 0.7, 7, and 18%, respectively, in 4 days from the conventional-till plot.
80	Majewski and others, 1995	Air	Oct-Nov 1992 (10 sampling days)	76	Salinas, CA	Methyl bromide	Field volatility experiment	4 $\mu g/m^3$	Injected 23 cm into soil. One field tarped with high density plastic film, the other left open to the air. Lost 22 and 89%, respectively, in 5 days by volatilization.

Table 2.3. Characteristics and summaries of state and local pesticide monitoring studies

[Letters under Occurrence refer to Location(s). Cont., continued; Dep, deposition; ND, not detected; NR, not reported; OA, oxygen analog transformation of the parent compound; PAHs, polycyclic aromatic hydrocarbons; PCBs, polychlorinated biphenyls; cm, centimeter; ft, feet; km, kilometer; <LOD, less than analytical limit of detection; m, meter; mi, miles; ng, nanogram; ng/m^3, nanogram per cubic meter; ng/smpl, nanogram per sample; pg, picogram; ppm, parts per million; >, greater than; <, less than; μg, microgram; μg/L, microgram per liter]

Study no.	Sampling				Compounds			Comments
	Study	Matrix	Date	Quantity	Location(s)	Name	Occurrence	Detection limits	
1	Cohen and Pinkerton, 1966	Dry-Dep. Rain	Jan 1965	NR	a Cincinnati, OH b Coshocton, OH	2,4-D (rain) 2,4,5-T Atrazine (rain) Chlordane DDE DDT Dieldrin Heptachlor epoxide Ronnel	b a b a a a a a a	NR	Collected dust deposited from a dust storm which originated in the Southern High Plains area and analyzed it for various halide and sulfur containing pesticides. Also collected rain three weeks after spraying an experimental plot with atrazine and 2,4-D. These samples were taken nearly 1 mi from the treated plot and concluded that the amounts found in rain were comparable to the amount sprayed.
2	Bevenue and others, 1972	Rain Snow	1971-72	NR	a Kailua, HI b Kaneohe, HI c Waipahu, HI d Honolulu, HI	Chlordane DDT, p,p'- Dieldrin HCH, γ-	All All All All	NR	Investigated the ambient concentrations in urban and urban/industrial areas. All compounds found at all locations but not in every sample.
3	Que Hee and others, 1975	Air	May-July 1972 (Daily sampling)	480	Sites in Saskatchewan, Canada: a Saskatoon b Naicam c Rosetown	Total 2,4-D 2,4-D esters: butyl octyl	All	0.1 ng 20 pg 40 pg	Implied that higher 2,4-D air concentrations occur from urban use than agricultural use. Sampled at one urban and two rural sites. Reported air concentrations in nanograms per cubic meter per day.

Table 2.3. Characteristics and summaries of state and local pesticide monitoring studies--*Continued*

| Study no. | Study | Sampling | | | | Compounds | | | Comments |
		Matrix	Date	Quan-tity	Location(s)	Name	Occurrence	Detection limits	
4	Young and others, 1976	Dry-Dep.	July-Sept 1973 Mar-June 1974 (1-week periods for 13 weeks)	NR	14 coastal and 6 island stations between Point Conception, CA and the Mexican border.	DDD, p,p'- DDE, p,p'- DDT, o,p'- DDT, p,p'-	All All All All	NR NR NR NR	Collected total dry deposition onto mineral oil coated glass plates. Analyzed samples and determined total DDT deposition and most probable sources.
5	Arthur and others, 1976	Air	1972-74 (Weekly com-posits of 4.3 minutes per hour for 7 days)	156	a Stoneville, MS	Aldrin DDD, o,p'- DDD, p,p'- DDE, o,p'- DDE, p,p'- DDT, o,p'- DDT, p,p'- DEF Diazinon Dieldrin Endrin HCH, α- HCH, γ- Heptachlor Heptachlor epoxide Malathion Methyl parathion Parathion Toxaphene	a a a a a a a a a a a a a a a a a a a	NR NR NR NR NR NR NR NR NR NR NR NR 0.1 ng/m^3 NR NR NR 1 ng/m^3 0.1 ng/m^3	Organophosphorus pesticides were found between June and Oct. Methyl parathion in June-Oct 1974 and Jan-May. DEF found only between Sept-Oct. In general, the highest concentrations were found during the summer months. DDT, methyl parathion and toxaphene were found most frequently and at the highest concentrations. Only reported maximum concentrations found for all but four compounds. Monthly averages for total DDT, methyl parathion, toxaphene, and endrin were presented.

Table 2.3. Characteristics and summaries of state and local pesticide monitoring studies—*Continued*

Study no.	Study	Sampling			Compounds			Comments
		Matrix	Date	Quan-tity	Location(s)	Name	Occurrence	Detection limits	

Study no.	Study	Matrix	Date	Quan-tity	Location(s)	Name	Occurrence	Detection limits	Comments
6	Farwell and others, 1976	Air	May-June 1973 April-June 1974 (24-hour samples daily at each site)	NR	a Prosser, WA (1973, 1974) b Kennewick, WA (1973, 1974) c Benton City, WA (1973, 1974) d Snake River, WA (1973, 1974) e Sunnyside, WA (1973) f Plymouth, WA (1973) g Mabton, WA (1973, 1974) h Hanford, WA (1973) i Burbank, WA (1974) j Travis, WA (1974) k Mercer, WA (1974)	4-D acid 2,4-D esters: 2-ethylhexyl butoxyethanol dimethylamine ethyl isobutyl isooctyl isopropyl methyl n-butyl PGBE (propylene-glycol butyl ether)	All. 1 or more high, medium and low volatile esters found at all sites.	NR <0.01 µg/m³ (inferred from data for all compounds)	Air sampled for high, low and medium volatility 2,4-D compounds in the grape growing region and the surrounding wheat growing area across a principal wind trajectory. This suggests a regional transport of the higher volatile esters, especially n-butyl ester. Presented averaged monthly concentrations only.

Table 2.3. Characteristics and summaries of state and local pesticide monitoring studies--*Continued*

Study no.	Study	Sampling			Compounds			Comments
		Matrix	Date	Quantity	Name	Occurrence	Detection limits	
7	Grover and others, 1976	Air	1966-68 1970-75	NR	2,4-D 2,4-D esters: isooctyl isopropyl methyl mixed butyl	Approximately 50% of all samples at all locations contained 1 or more 2,4-D esters	$0.001\ \mu g/m^3$ $0.05\ \mu g/m^3$ ND $0.001\ \mu g/m^3$ $0.001\ \mu g/m^3$	50% of the samples contained 2,4-D. The most frequently found ester was the butyl. The higher volatile esters were found in the highest concentrations and the highest proportions in air.
					Sites in Saskatchewan, Canada: a Aberdeen (1970) b Regina (1966-75) c Saskatoon (1971-75) d Indian Head (1972-73) e Naicam (1971-73) f North Battleford (1973) g Rosetown (1971-73) h Swiftcurrent (1972-73) i Yorkton (1973)					
8	Munson, 1976	Air Rain	July-Nov 1974	NR	Toxaphene	a, b, c	NR	Rainwater collected as part of a study to determine residues in rainwater, air, ground water, storm sewer water, suspended sediments, bottom sediments, and zooplankton in, or near the Chesapeake Bay.
					a Fort Smallwood, MD b Sollers Point, MD c Morrell Park, MD			
9	Peakall, 1976	Rain Snow	Aug 1974 Aug 1975	NR	DDT, total	Aug 2, 1974- Dec 18 1974	0.1 ng/L	Collected 2-3 week composite precipitation samples for 1 year. DDT residues thought to have originated in WA.
					Ithaca, NY			

Table 2.3. Characteristics and summaries of state and local pesticide monitoring studies--*Continued*

Study no.	Study	Sampling			Compounds			Comments
		Matrix	Date	Quantity	Location(s)	Name	Occurrence	Detection limits	
10	Reisinger and Robinson, 1976	Air	Apr-June 1973 Apr-June 1974	NR	a Prosser, WA (1973, 1974) b Kennewick, WA (1973, 1974) c Benton City, WA (1973, 1974) d Snake River, WA (1973, 1974) e Mabton, WA (1973, 1974) f Burbank, WA (1974) g Travis, WA (1974) h Mercer, WA (1974)	2,4-D	All	NR	Investigated the meteorological conditions to explain high air concentrations and drift of 2,4-D and related esters.
11	Giam and others, 1980	Air	Mar-Apr 1977 (24-hour samples, 10 sites)	NR	10 sites in the northwest Gulf of Mexico	DDT, *p,p'*- DDE, *p,p'*-	All All All	NR	Analyzed air samples for DDTs in conjunction with a survey of PCB and phthalate ester analysis.
12	Harder and others, 1980	Air Rain	July 1977- Dec 1978	9 17	4 sites in an estuary near Georgetown, SC	DDE, *p,p'*- DDT, *p,p'*- Toxaphene	All All All	<0.1 ng/L <0.1 ng/L <8 ng/L, 15 ng	Monitored toxaphene input into a saltwater marsh estuary. Found the highest concentrations began in late spring and corresponded to its use. Maximum concentrations observed from June-Aug and Sept.-Oct. Reported air concentrations in nanograms per kilogram.

Table 2.3. Characteristics and summaries of state and local pesticide monitoring studies--*Continued*

Study no.	Study	Sampling				Compounds			Comments
		Matrix	Date	Quantity	Location(s)	Name	Occurrence	Detection limits	
13	Eisenreich and others, 1981	Air Rain	NR	NR	Great Lakes. Specific sites not provided.	DDT, total Dieldrin Endosulfan I Endosulfan II HCH, α- HCH, γ- Methoxychlor	Throughout the Great Lakes region	NR	Main focus was on PCBs in the Great Lakes.
14	Grover and others, 1981	Air	1978, 1979 (May-Nov)	Daily	Sites in Saskatchewan, Canada: a Regina b Indian Head (1979 only)	Triallate	All	0.5 ng/m^3	Found the highest concentrations during both spring and fall application periods. Samplers located in a high use area.
15	Wu, 1981	Rain Snow Dry-Dep.	Dec 1976-Feb 1979	68	Chesapeake Bay near Annapolis, MD	Atrazine	Year round	60 pg 2 ng/L (inferred from data)	Analyzed rainwater throughout the year.
16	Rapaport and others, 1985	Rain Snow Peat bog cores	1981-84	NR	a Alfred, Ontario, Canada b Big Heath, ME c Diamond, Ontario, Canada d Marcell, MN e Fourchu, Nova Scotia f Minneapolis, MN	DDT, *p,p'*-DDT, total	All All	NR 0.05 ng/L DDE	Analyzed peat bog cores with some rain and snow samples. Reported detection limit for total DDT in terms of DDE only.

Table 2.3. Characteristics and summaries of state and local pesticide monitoring studies--*Continued*

Study no.	Study	Sampling				Compounds			Comments
		Matrix	Date	Quantity	Location(s)	Name	Occurrence	Detection limits	
17	Rapaport and Eisenrich, 1986, 1988	Peat bog cores	Spring-summer (1981-84, both studies)	NR	a Marcell, MN b Diamond, Ontario, Canada c Alfred, Ontario, Canada d Lac St. Jean, Quebec, Canada e Croatan, NC f Big Heath, ME g Fourchu, Nova Scotia	DDT (1988) HCB (1988) HCHs (1988) Toxaphene (1986, 1988)	All All All but f All	NR	Reported DDT and toxaphene concentrations in air, rain, and snow from some direct measurements and from peat core concentrations. Both reports used data generated from the same sampling sites and times.
18	Sava, 1985	Air	June 1985	NR	3 sites in residential Salinas, CA: 1, 2, and 3	Chlorpyrifos Dacthal DDD, *p,p'*- DDE, *p,p'*- Diazinon Endosulfan I Endosulfan II Furadan Malathion Methomyl Methoxychlor Methyl parathion Parathion Phosdrin Tetradifon	<LOD <LOD <LOD <LOD <LOD sites 1, 2 <LOD <LOD <LOD <LOD <LOD site 1 site1 site 3 <LOD	0.009 μg/m³ 0.017 μg/m³ 0.017 μg/m³ 0.017 μg/m³ 0.005 μg/m³ 0.009 μg/m³ 0.017 μg/m³ 0.024 μg/m³ 0.017 μg/m³ 0.035 μg/m³ 0.069 μg/m³ 0.007 μg/m³ 0.017 μg/m³ 0.010 μg/m³ 0.052 μg/m³	Analyzed ambient air in an urban area near an agricultural area during a high pesticide use month.

Table 2.3. Characteristics and summaries of state and local pesticide monitoring studies--*Continued*

Study no.	Study	Sampling				Compounds			Comments
		Matrix	Date	Quan-tity	Location(s)	Name	Occurrence	Detection limits	
19	Strachan, 1985	Rain	May-Oct 1983 Apr-Nov 1983	3 5	a Isle Royale, Lake Superior b Caribou Island, Lake Superior	Chlordane, *cis-*	ND	1.0 ng/smpl	Locations at Caribou Island (East Lake Superior) were lower in concentrations than the Isle Royale concentrations. Concentrations at both locations were higher during the spring and summer.
						Chlordane, *trans-*	ND	1.0 ng/smpl	
						DDD, *p,p'-*	a, b	0.2-1 ng/smpl	
						DDE, *p,p'-*	a, b	0.2-1 ng/smpl	
						DDT, *o,p'-*	ND	0.4 ng/smpl	
						DDT, *p,p'-*	a, b	0.2-1 ng/smpl	
						Dieldrin	a, b	0.2-1 ng/smpl	
						Endrin	a, b	0.2-1 ng/smpl	
						HCB	a, b	0.2-1 ng/smpl	
						HCH, α-	a, b	1-5 ng/smpl	
						HCH, γ-	a, b	1-5 ng/smpl	
						Heptachlor epoxide	a, b	0.2-1 ng/smpl	
						Methoxychlor	a, b	1-5 ng/smpl	
						Mirex	ND	0.4 ng/smpl	
						Toxaphene	ND	5.0 and 50 ng/smpl	

Table 2.3. Characteristics and summaries of state and local pesticide monitoring studies--*Continued*

Study no.	Sampling					Compounds			Comments
	Study	Matrix	Date	Quan-tity	Location(s)	Name	Occurrence	Detection limits	
20	Glotfelty and others, 1987	Air Fog	May 1984 and Jan 1985	NR	a Parlier, CA b Corcoran, CA c Lodi, CA d Beltsville, MD	Alachlor	d	NR	Sampled during winter months only in cotton, orchard, and grape growing areas.
						Atrazine	All		
						Chlorpyrifos	a, b, c		
						Chlorpyrifos OA	b, c		
						DEF	a, c		
						Diazinon	All		
						Diazinon OA	c		
						Malathion	All		
						Methidathion	a, b, c		
						Methidathion OA	b, c		
						Methyl parathion	d		
						Metolachlor	d		
						Parathion	a, b, c		
						Parathion OA	a, b, c		
						Pendimethalin	a, c		
						Simazine	All		

Table 2.3. Characteristics and summaries of state and local pesticide monitoring studies--*Continued*

Study no.	Study	Sampling				Compounds			Comments
		Matrix	Date	Quantity	Location(s)	Name	Occurrence	Detection limits	
21	Shulters and others, 1987	Rain	Dec 1981- Mar 1983	NR	Sites in Fresno, CA: 1 Industrial site 1 Residential site	2,4-D	All	0.01 µg/L	Characterized pesticides in rain in two sites in Fresno, CA. Found that highest concentrations coincided with local use.
						2,4-DP	<LOD	0.01 µg/L	
						2,4,5-T	<LOD	0.01 µg/L	
						Aldrin	<LOD	0.01 µg/L	
						Chlordane	All	0.01 µg/L	
						DDD	<LOD	0.01 µg/L	
						DDE	<LOD	0.01 µg/L	
						DDT	<LOD	0.01 µg/L	
						Diazinon	All	0.005 µg/L	
						Dieldrin	<LOD	0.01 µg/L	
						Endosulfan	All	0.01 µg/L	
						Endrin	<LOD	0.01 µg/L	
						Ethion	<LOD	0.01 µg/L	
						HCH, γ-	All	0.01 µg/L	
						Malathion	All	0.01 µg/L	
						Methyl parathion	<LOD	0.01 µg/L	
						Methyl trithion	<LOD	0.01 µg/L	
						Methoxychlor	All	0.01 µg/L	
						Parathion	All	0.01 µg/L	
						Silvex	<LOD	0.01 µg/L	
						Trithion	<LOD	0.01 µg/L	
22	Grover and others, 1988a	Air	1981 and 1982 (24-hour weekday, 72-hour weekend sampling from May-Nov)	NR	Sites in Saskatchewan, Canada: a Regina (1981) b Melfort (1981, 1982)	Triallate	All	0.05 ng/m^3	Maximum air concentrations corresponded to rainfall events. Lowest ambient air concentrations during dry seasons. Higher ambient concentrations during wet years than dry ones.
						Trifluralin	All	0.05 ng/m^3	

Table 2.3. Characteristics and summaries of state and local pesticide monitoring studies—*Continued*

Study no.	Sampling				Compounds			Comments
	Study	Matrix	Date	Quantity	Location(s)	Name	Occurrence	Detection limits	
23	Oudiz and Klein, 1988	Air	Jan-Feb 1986 Sept-Oct, 1986 (24-hour samples, 4 times per week each site)	NR	North San Joaquin Valley: a Sanger, CA b Parlier-1, CA c Parlier-2, CA d Reedley, CA e Dinuba, CA f Selma, CA South San Joaquin Valley: g Shafter, CA h Wasco, CA i McFarland CA j Delano-1, CA k Delano-2, CA l Earlimart, CA Imperial Valley: m Heber, CA n Holtville, CA o Brawley-1, CA p Brawley-2, CA q Calipatria-1, CA r Calipatria-2, CA s El Centro (background site)	Parathion	North San Joaquin Valley: All South San Joaquin Valley: h, i, j, l Imperial Valley: All	0.01 µg/m^3	Ambient air sampling done in conjunction with a study on the ambient levels of methyl parathion as a toxic air component. Highest air concentrations were measured in the northern San Joaquin Valley. Lower concentrations were measured in the southern San Joaquin Valley and Imperial Valley. The highest concentrations seemed to be correlated to local use.
						Parathion OA	North San Joaquin Valley: b, d, e South San Joaquin Valley: None Imperial Valley: None	0.02 µg/m^3	

Table 2.3. Characteristics and summaries of state and local pesticide monitoring studies--*Continued*

Study no.	Study	Sampling			Location(s)	Compounds			Comments
		Matrix	Date	Quantity		Name	Occurrence	Detection limits	
24	Seiber and others, 1989	Air	May-June 1986 (4 times weekly for 5 weeks)	NR	a Davis, CA (background site) b Maxwell, CA c Robbins, CA d Trowbridge, CA e Williams, CA	Methyl parathion Methyl parathion OA Molinate Thiobencarb	b, c, d, e b, e b, c, d, e b, c, d, e	0.2 ng/m^3 0.5 ng/m^3 1.4 ng/m^3 2.0 ng/m^3	Air samples taken in urban areas surrounded by application sites. Found a correlation between air concentrations and use.
25	Turner and others, 1989	Fog	Jan 1986	NR	Northcentral Stanislaus County, CA (3 sites)	Chlorpyrifos Diazinon Methidathion Parathion	All All All All	For all compounds: 1,000 ng/L	Investigated spray and volatilization drift and deposition on nontarget crops by fog.
26	Glotfelty and others, 1990c	Air Rain	Apr 1981-Aug 1982 (7 sites, rain; 2 sites, air)	NR	a Wye Institute, MD b Wye Plantation, MD c Patuxent Wildlife Center, MD d Sandy Point State Park, MD	Alachlor Atrazine Metolachlor Simazine Toxaphene	All All All All All	0.005 µg/L and ng/m^3 0.001 µg/L 0.005 µg/L and ng/m^3 0.001 µg/L 0.005 µg/L	Year around detection of triazines with seasonal increases coinciding with local use. Implications for transport into the area from the southern growing areas before local use begins.
27	Muir and others, 1990	Rain	May-Oct 1989	NR	Experimental Lakes Area near Kenora, Ontario, Canada	Atrazine Bromoxinil Dacthal Dicamba Diclofop Triallate Trifluralin	All All NR All 2 samples All All	NR	Herbicides detected at >250 km from nearest agricultural region. Atrazine and Dacthal are not in major use within 300 km. Study implications of long-range transport.

Table 2.3. Characteristics and summaries of state and local pesticide monitoring studies--Continued

Study no.	Study	Sampling				Compounds			Comments
		Matrix	Date	Quantity	Location(s)	Name	Occurrence	Detection limits	
28	Brun and others, 1991	Rain	1980-89 (Monthly composite)	NR	a Kejimkujik National Park, Nova Scotia b Ellerslie, Prince Edward Island, Canada c Jackson, Nova Scotia	Aldrin	**	0.01 µg/L	Detected heptachlor, cis-chlordane, endosulfan I and II, and aldrin occasionally at low levels. DDT was found only two times at Ellerslie. α-HCH and γ-HCH detections were the most abundant with the highest concentrations measured between 1980-84. Seasonal patterns noted with the highest concentrations occurring during the spring and fall. Study done in conjunction with PCBs, PAHs, and chlorinated benzenes. Data was presented in graphic form only. No tabular data showed actual concentrations.
						Chlordane, cis-	**	0.005 µg/L	
						Chlordane, trans-	**	0.005 µg/L	
						DDD, p,p'-	b	0.001 µg/L	
						DDE, p,p'-	b	0.001 µg/L	
						DDT, o,p'-	b	0.001 µg/L	
						DDT, p,p'-	b	0.001 µg/L	
						Dieldrin	**	0.001 µg/L	
						Endosulfan I	**	0.01 µg/L	
						Endosulfan II	**	0.01 µg/L	
						Endrin	**	0.01 µg/L	
						HCH, α-	All	0.01 µg/L	
						HCH, γ-	All	0.01 µg/L	
						Heptachlor	**	0.01 µg/L	
						Heptachlor epoxide	**	0.01 µg/L	
						Methoxychlor, p,p'-	**	0.01 µg/L	
						Mirex	** (**, detected sporadically)	0.001 µg/L	
29	Capel, 1991	Rain Snow	1989-90	NR	a Rosemont, MN b St. Paul, MN	Alachlor	All	20 ng/L	Found occurrence to be seasonal mainly during March-May. Maximum concentrations occurred following applications. Atrazine detected in rain and snow throughout the year. Alachlor and cyanazine <LOD after July.
						Atrazine	All	20 ng/L	
						Cyanazine	All	20 ng/L	

Table 2.3. Characteristics and summaries of state and local pesticide monitoring studies--*Continued*

Study no.	Study	Sampling				Compounds			Comments
		Matrix	Date	Quan-tity	Location(s)	Name	Occurrence	Detection limits	
30	Knap and Binkley, 1991	Air	1985 (4 samples) 1986 (4 samples)	NR	a Adirondack Mountains, NY near Whiteface Mountain b Western Atlantic Ocean near Norfolk, VA	Chlordane (*cis+trans*) Dieldrin HCB HCH, α- HCH, γ-	a, b a, b a, b a, b a, b	NR <0.001 ng/m³ (inferred from data)	Determined the extent of specific compound distribution in the troposphere. Air samples taken at high (8,000 to 10,000 ft) and low (300 to 3,000 ft) altitudes.
31	Segawa and others, 1991	Air	Feb-Apr 1990	266	a Brea/La Habra, CA b Garden Grove, CA c San Fernando, CA d Rosemead/ Monrovia, CA	Malathion Malathion OA	All All	0.1 µg/smpl = (0.069 ng/m³ for high volume samples)	Monitored indoor and outdoor air concentrations before, during, and after an aerial application of malathion to an urban area. Also measured the OA.

Table 2.3. Characteristics and summaries of state and local pesticide monitoring studies—*Continued*

Study no.	Study	Sampling			Location(s)	Compounds			Comments
		Matrix	Date	Quantity		Name	Occurrence	Detection limits	
32	Hoff and others, 1992	Air	July 1988 - Sept 1989 (48-hour samples to July 1, 1989; 24-hour samples every 6th day thereafter)	143	a Egbert, Ontario, Canada	Chlordane, *cis*- Chlordane, *trans*-	All All	0.04-0.1 pg/m^3 (only a range given for all compounds)	A year-long study measuring the annual cycling of these compounds plus PCBs in the atmosphere. The highest concentrations were found between May and Sept.
						DDD, *p,p'*-	All		
						DDE, *p,p'*-	All		
						DDT, *o,p'*-	All		
						DDT, *p,p'*-	All		
						Dieldrin	All		
						Endosulfan II	All		
						Endrin	All		
						HCH, α-	All		
						HCH, β-	All		
						HCH, γ-	All		
						Heptachlor	All		
						Heptachlor epoxide	All		
						Methoxychlor	All (except Nov-Dec)		
						Mirex	5 samples only		
						Nonachlor, *cis*-	All		
						Nonachlor, *trans*-	All		
						Oxychlordane	All		
						Toxaphene	All (except Jan-Feb)		
						Trifluralin	All		
							Average monthly detections.		

Table 2.3. Characteristics and summaries of state and local pesticide monitoring studies--*Continued*

| Study no. | Sampling | | | | Compounds | | | Comments |
	Study	Matrix	Date	Quan-tity	Location(s)	Name	Occurrence	Detection limits	
33	Lane and others, 1992	Air	1985-89 (24-, 40- and 44-hour sampling periods)	82	Sites in Ontario, Canada: a Mississauga b Stoney Creek c Niagara-on-the-Lake d Niagara-on-the-Lake (offshore) e Turkey Lake f Pointe Petre g Sandbanks Provincial Park	HCB HCH, α- HCH, γ-	All All All	7 pg/m^3 14 pg/m^3 15 pg/m^3	Sampled air for HCHs and HCB. Found relatively little of each compound on particulate matter. The HCH concentrations increased during the summer and were lowest during the winter. HCB concentrations remained fairly constant throughout the year.

Table 2.3. Characteristics and summaries of state and local pesticide monitoring studies--*Continued*

Study no.	Study	Sampling Matrix	Date	Quantity	Location(s)	Compounds Name	Occurrence	Detection limits	Comments
34	Nations and Hallberg, 1992	Rain Snow	Oct 1987-Sept 1990 (Event composite and several sequential samples)	325	a Big Springs basin, IA b Bluegrass water shed, IA c Iowa City, IA	Alachlor	All	For all compounds: 0.1 µg/L	Implications of regional transport into area. Urban and rural sampling areas. Found occurrences to be seasonally distributed. No pesticides detected in snow. (a) One site located near row–crop area at 2 m elevation. One site located 11 km away in a no pesticide use area at ground level. (b) One site at 11 m elevation and one site at ground level. Both sites within 2 km of each other and in an urban setting. (c) Both sites at ground level near row-cropped areas and about 10 km apart.
						Atrazine	All		
						Butylate	All		
						Chlorpyrifos	<LOD		
						Cyanazine	All		
						Diazinon	<LOD		
						Dimethoate	All		
						EPTC	All		
						Ethoprop	<LOD		
						Fonofos	All		
						Malathion	All		
						Methyl Parathion	All		
						Metolachlor	All		
						Metribuzin	All		
						Parathion	<LOD		
						Pendimethalin	All		
						Phorate	<LOD		
						Propachlor	All		
						Terbufos	<LOD		
						Trifluralin	All		
35	Zabik and Seiber, 1992	Air Rain Snow Fog	Dec 90-Mar 91	NR	a Lindcove Field Station, CA Sequoia National Park, CA: b Ash Mountain Station c Lower Kaweah Station	Chlorpyrifos	Air: a, b; rain: a, b, c	For all compounds: 0.7 pg/m^3; 1.3 pg/L	Measured air and wet deposition along an elevation gradient (114-1,920 m). Aerial transport of pesticides from high use areas into mountain region. One fog sample taken at the lowest site.
						Diazinon	Air: a, b; rain: a, b, c		
						Diazinon OA	Air: a, b; rain: a, b, c		
						Parathion	Air: a, b; rain: a, b, c		
						Parathion OA	All		

Table 2.3. Characteristics and summaries of state and local pesticide monitoring studies--*Continued*

Study no.	Sampling				Compounds			Comments
	Study	Matrix	Date	Quantity	Location(s)	Name	Occurrence	Detection limits	
36	Brown and others, 1993	Air	May 1990 (3 sites for 10 days)	NR	Garden Grove, CA	Malathion Malathion OA		NR	Monitored air concentrations before, during and after an aerial application of malathion to an urban area. Also measured the OA and other metabolites.
37	Seiber and others, 1993	Air Fog	Jan 1989 Jan 15-25, 1989	NR	Parlier, CA	Chlorpyrifos	All	1.0, 1.6, 2.4 ng/m^3	Took 24-hour air samples along with 12-hour day and 12-hour night samples. Also sampled night-time fog and air when fog occurred. No LOD for fog water samples reported.
						Chlorpyrifos OA	Most samples	2.0, 3.2, 4.9 ng/m^3	
						Diazinon	All	1.0, 1.6, 2.4 ng/m^3	
						Diazinon OA	Many samples	2.0, 3.2, 4.9 ng/m^3	
						Methidathion	All	1.0, 1.6, 2.4 ng/m^3	
						Methidathion OA	Many samples	2.0, 3.2, 4.9 ng/m^3	
						Parathion	All	1.0, 1.6, 2.4 ng/m^3	
						Parathion OA	Many samples	2.0, 3.2, 4.9 ng/m^3	

Table 2.3. Characteristics and summaries of state and local pesticide monitoring studies--*Continued*

Study no.	Study	Sampling				Compounds			Comments
		Matrix	Date	Quan-tity	Location(s)	Name	Occurrence	Detection limits	
38	Muir and others, 1993	Lichen	Sept-Oct 1986 (30 sites) Sept-Oct 1987 (15 sites) Triplicate samples.	NR	Sites in Ontario, Canada: a Experimental Lakes Area b Kakabeka c Ouimet Canyon d Sibley e Beardmore f Rossport g Ripple h White Lake i Wawa j Agawa k Turkey Lakes l Dunns Valley m Chapleau n Elliot Lake o Wanhoe Lake p Remi Lake q Manitoulin	Chlordanes (*cis*- and *trans*-chlordane, *cis*- and *trans*-nonachlor, oxychlordane, and heptachlor epoxide) DDTs (*p,p'*-DDE, -DDD, -DDT, and *o,p'*-DDT) Dieldrin Endosulfan HCB HCHs (α+γ) Toxaphene	1986-87 1986-87 1986-87 1987 1986-87 1986-87 1986-87	For all compounds: 0.02-0.05 ng/g (dry weight)	Sampled lichens and analyzed them for various organochlorine compounds. Determined that concentrations of DDTs, chlordanes, and dieldrin were significantly higher in south-central Ontario. HCHs had higher concentrations at the northern and northwestern sites.

Table 2.3. Characteristics and summaries of state and local pesticide monitoring studies—*Continued*

Study no.	Study	Sampling				Compounds			Comments
		Matrix	Date	Quantity	Location(s)	Name	Occurrence	Detection limits	
38--Cont.	Muir and others, 1993	Lichen	Sept-Oct 1986 (30 sites) Sept-Oct 1987 (15 sites) Triplicate samples.	NR	Sites in Ontario, Canada (cont): r Espanola s Killarney t Windy Lake u Point au Baril v Luther Bog w Burt Lake x Cache Bay y Parry Sound z Mactier aa Bear Lake ab North Bay ac Gravenhurst ad Dorset ae Haliburton af Whitney ag Bancroft ah Petawawa ai Charleston	Chlordanes (*cis*- and *trans*-chlordane, *cis*- and *trans*-nonachlor, oxychlordane, and heptachlor epoxide) DDTs (*p,p'*-DDE, -DDD, -DDT, and *o,p'*-DDT) Dieldrin Endosulfan HCB HCHs (α+γ) Toxaphene	1986-87 1986-87 1986-87 1987 1986-87 1986-87 1986-87	For all compounds: 0.02-0.05 ng/g (dry weight)	Sampled lichens and analyzed them for various organochlorine compounds. Determined that concentrations of DDTs, chlordanes, and dieldrin were significantly higher in south-central Ontario. HCHs had higher concentrations at the northern and northwestern sites.

Table 2.4. Characteristics and summaries of national and multistate pesticide monitoring studies

[Letters under Occurrence refer to Location(s). Cont, continued; Dep, deposition; ND, not detected; NR, not reported; OA, oxygen analog transformation of the parent compound; PCBs, polychlorinated biphenyls; ptcl, particulate matter; cm, centimeter; ft, feet; km, kilometer; <LOD, less than analytical limit of detection; m, meter; mi, miles; ng, nanogram; ng/m^3, nanogram per cubic meter; ng/smpl, nanogram per sample; pg, picogram; ppm, parts per million; >, greater than; <, less than; μg, microgram; μg/L, microgram per liter]

Study no.	Study	Sampling				Compounds				Comments
		Matrix	Date	Quan-tity	Location(s)	Name	Occurrence	Detection limits		
1	Tabor, 1965	Air	1963:			Aldrin	h	For all compounds:		An investigation of human exposure to pesticides in air. This study investigated the airborne particle fraction only due to sampling limitations. The urban areas sampled were (1) near agricultural areas where pesticides were applied to crops and susceptible to drift, or (2) areas with local insect control applications. Not all pesticides analyzed for were used at all sampling locations.
			May-June	5	a Edinburg, TX	Chlordane	d, e, f, g, i	<0.1 ng/m^3		
			May-June	5	b Weslaco, TX	DDT	all but c, i, k	(inferred from data)		
			July	2	c Harlington, TX	Malathion	j, k, l, m			
			1964:			Toxaphene	g			
			May-June	16	d Fort Valley, GA					
			May-June	8	e Inman, SC					
			July-Sept	6	f Leland, MS					
			July-Sept	5	g Newellton, LA					
			Oct-Dec	6	h Florida City, FL					
			June-Aug	6	i Lake Alfred, FL					
			July	3	j Houston, TX					
			Aug-Sept	7	k A Chesapeake Bay Community					
			Sept	12	l An Atlantic Coast resort community					
			Aug-Sept	11	m A large eastern city					

Table 2.4. Characteristics and summaries of national and multistate pesticide monitoring studies--*Continued*

Study no.	Study	Sampling			Location(s)	Compounds			Comments
		Matrix	Date	Quantity		Name	Occurrence	Detection limits	
2	Stanley and others, 1971	Air	1967-68: (Two 12-hour or one 24-hour sample per day at each site.)	880	a Baltimore, MD b Buffalo, NY c Dothan, AL d Fresno, CA e Iowa City, IA d Orlando, FL g Riverside, CA h Salt Lake City, UT i Stoneville, MS	2,4-D and esters Aldrin Chlordane DDD, *p,p'*- DDE, *o,p'*- DDE, *p,p'*- DDT, *o,p'*- DDT, *p,p'*- DEF Dieldrin Endrin HCH, α- HCH, β- HCH, δ- HCH, γ- Heptachlor Heptachlor epoxide Malathion Methyl parathion Parathion Toxaphene	h e ND ND c, f, i a, c, d, e, f, g, i All All i f i a, d, e, h a, h h a, e, h e,f ND f c, f, i f c, f, i	For all compounds: 0.1 ng/m³	Found only *p,p'*- and *o,p'*-DDT at all locations. Other compounds were found at different times of the year. The highest levels occurred in the agricultural areas of the southern United States. Reported only highest levels found in the 9 cities for the 18 compounds. Also, only reported concentration ranges for 6 compounds in Stoneville, MS and 3 compounds in Orlando, FL.

Table 2.4. Characteristics and summaries of national and multistate pesticide monitoring studies--*Continued*

Study no.	Study	Sampling			Location(s)	Compounds		Detection limits	Comments
		Matrix	Date	Quantity		Name	Occurrence		
3	Kutz and others, 1976	Air	1970-72 (30 samples; two 24-hour samples per day at each site)	2,479	a Montgomery, AL b Little Rock, AR c Normal, IL d Topeka, KS e Louisville, KY (1972) f Peaks Mill, KY g Frankfort, KY (1971) h Monroe, LA (1972) i Alexandria, LA j Augusta, ME k Yellow Bay, MT (1972) l Helena, MT m Billings, MT (1971) n San Miguel, NM (1971-72) o Albuquerque, NM (1971) p Macedonia, NC (1971-72) q Meadows, NC r Columbus, OH (1972) s London, OH t Tulsa, OK (1972)	Aldrin Azodrin Chlordane Dachtal DDD, *o,p'-* DDD, *p,p'-* DDE, *o,p'-* DDE, *p,p'-* DDT, *o,p'-* DDT, *p,p-* Diazinon Dieldrin Disyston* Endosulfan I Endosulfan II Endrin Folex HCH, α- HCH, β-* HCH, γ- (*Sampled in 1975 only.)	All but n,o b, n, o ac, ae, l, v, w c, d, n, t e, f, g All but c, j, n, o, u ND All All All All but a All but ac ac, ad, ae b, c, d, l, q b a, b, f, aa, ab, ad aa All ad All	For all compounds: 1-10 ng/m^3 (range only reported)	Part of a national pesticide monitoring program. Air samples taken at sites selected for their potential of having high ambient air concentrations. Sampled suburban air in 1975.

Table 2.4. Characteristics and summaries of national and multistate pesticide monitoring studies—*Continued*

Study no.	Study	Sampling			Location(s)	Compounds			Comments
		Matrix	Date	Quantity		Name	Occurrence	Detection limits	
3—Cont.	Kutz and others, 1976	Air	1970-72: (30 samples; two 24-hour samples per day at each site)	2,479	u Chickasha, OK v Portland, OR (1971-72) w Salem, OR x Camp Hill, PA (1972) y Gettysburg, PA z Pierre, SD aa Memphis, TN ab Lucy, TN (1972) ac Miami, FL (1975)* ad Jackson, MS (1975)* ae Fort Collins, CO (1975)* (*Sampled in 1975 only.)	Heptachlor Heptachlor epoxide* Kelthane Leptophos Malathion Methyl parathion Oxychlordane* Parathion Phorate Toxaphene Trifluralin 2,4-D Butyl 2,4-D Butoxyethyl 2,4-D Isopropyl 2,4,5-T BOEE 2,4,5-T Isooctyl	All but ae ad g u All All but x, y, ac, ad, ae ac, ad ac, y, u, a, b, o, p, q aa, v, b, j, n, o a, b, h, i, u, aa, ab All but a, k, l, m, v, w, x, y, z All d, h, p, z All but a, e, f, g, j, p, q, x, y, aa, ab c, aa aa	For all compounds: 1-10 ng/m^3 (range only reported)	Part of a national pesticide monitoring program. Air samples taken at sites selected for their potential of having high ambient air concentrations. Sampled suburban air in 1975.

Table 2.4. Characteristics and summaries of national and multistate pesticide monitoring studies—*Continued*

Study no.	Study	Sampling				Compounds			Comments
		Matrix	Date	Quantity	Location(s)	Name	Occurrence	Detection limits	
4	Strachan and Huneault, 1979	Rain Snow	1975-76 May-Nov, 1976	NR	All sites in Ontario, Canada: Rain: 7 sites adjacent to the Great Lakes a Sibley b Batchawana Bay c South Baymouth d Goderich e Pelle Island f CCIW, near Burlington g Picton Snow: 17 sites downwind of the Great Lakes h Nipgon i Geralton j Marathon k Wawa l Hearst b Batchawana Bay m Thessalon n Espanola o Parry Sound p Honey Harbor q Orangeville r Peterborough s Ivy Lee t Ottawa u Iroquois Falls v Marten River w Trout Creek	Aldrin DDTs Dieldrin Endosulfan I Endosulfan II Endrin HCH, α- HCH, γ- Heptachlor Heptachlor epoxide Methoxychlor Mirex	Present throughout the Great Lakes region	1-2 ng/L 1 ng/L 1 ng/L 1 ng/L 1 ng/L 1 ng/L 2 ng/L 2 ng/L 1-2 ng/L 1-2 ng/L 2 ng/L 1-2 ng/L	Detected lower concentrations in snow than in rain, except for PCBs. The highest concentrations were found in rain during the midsummer months. Snow concentrations were lower than in rain except for the PCBs. Endosulfan I and II, and dieldrin were generally not found in snow in one area.

Table 2.4. Characteristics and summaries of national and multistate pesticide monitoring studies—*Continued*

Study no.	Study	Sampling				Compounds			Comments
		Matrix	Date	Quan-tity	Location(s)	Name	Occurrence	Detection limits	
5	Strachan and others, 1980	Rain	May-Nov, 1976, 1977	81	All sites in Ontario, Canada:	Aldrin	Present throughout the Great Lakes region.	For all compounds:	Collected rain and snow core samples at various sites around the Great Lakes and analyzed them for a variety of organo-chlorine compounds. Also examined the concentrations of these compounds associated with particulate matter during rainfall.
		Snow	Feb, 1976	17	Rain: 1976	Chlordane		1-2 ng/L	
					a Sibley	DDTs			
					b Batchawana Bay	Dieldrin	Chlordane,		
					c South Baymouth	Endosulfan I	endrin,		
					d Goderich	Endosulfan II	and HCB		
					e Pelle Island	Endrin	were not		
					f CCIW	HCB	detected		
					g Picton	HCH, α-	frequently.		
					Rain: 1977	HCH, γ-			
					h Upsala	Heptachlor			
					i Nagagamis Province Park	Heptachlor epoxide			
					j Batchawana Bay	Methoxychlor			
					f CCIW	Mirex			
					j Chalk River				
					k Picton				
					l Carillon Province Park				
					Snow:				
					m Nipgon				
					n Geralton				
					o Marathon				
					p Wawa				
					q Hearst				
					b Batchawana Bay				
					r Thessalon				
					s Espanola				
					t Parry Sound				
					u Honey Harbor				
					v Orangeville				
					w Peterborough				
					x Ivy Lee				
					y Ottawa				
					z Iroquois Falls				
					aa Marten River				
					bb Trout Creek				

Table 2.4. Characteristics and summaries of national and multistate pesticide monitoring studies--*Continued*

Study no.	Study	Matrix	Sampling Date	Quantity	Location(s)	Compounds Name	Occurrence	Detection limits	Comments
6	Rice and others, 1986	Air	Summer-Fall 1981 (24 hour samples for four 2-week periods each site)	NR	a Greenville, MS b St. Louis, MO c Bridgeman, MI d Beaver Isle, MI	Toxaphene	All	0.06 mg/m^3	Measured an increasing air concentration gradient from the north to the south. The results imply that there is a northward movement of pesticides via the atmosphere from major use areas.
7	Richards and others, 1987	Rain	Spring-Summer, 1985	79	a West Lafayette, IN b Tiffin, OH c Parsons, WV d Potsdam, NY	Alachlor Atrazine Butylate Carbofuran Cyanazine DDT Dieldrin EPTC Fonofos Linuron Metolachlor Metribuzin Pendimethalin Simazine Terbufos Toxaphene	All All a, b All All ND ND a a, b ND a, b, d b, d b All ND ND	0.1 µg/L 0.05 µg/L 0.05 µg/L 0.2 µg/L 0.25 µg/L NR NR 0.05 µg/L 0.05 µg/L 1.5 µg/L 0.25 µg/L 0.1 µg/L 0.05 µg/L 0.25 µg/L 0.1 µg/L NR	79 samples analyzed for 19 compounds, 11 detected in rain. Only reported if each compound was detected at various ranges. No actual highs, lows, or mean values were reported. Found that the highest concentrations coincided with spring/summer application times. Low concentrations found during fall/winter periods.

Table 2.4. Characteristics and summaries of national and multistate pesticide monitoring studies--*Continued*

Study no.	Study	Sampling				Compounds			Comments
		Matrix	Date	Quantity	Location(s)	Name	Occurrence	Detection limits	
8	Strachan, 1988	Rain	May-Oct 1984	NR	a Caribou Island, Lake Superior	Chlordanes (*cis+trans*)	ND	0.02 ng/L	The same compounds were found at all of the sites even though some were separated by >3,500 km. This implies that the source inputs were from different air masses. This indicates a global distribution for these types of pesticides.
					b Agawa Bay	DDD, p,p'-	a, c, d	0.02 ng/L	
					c Cree Lake,	DDE, p,p'-	All	0.02 ng/L	
					Saskatchewan, Canada	DDT, o,p'-	ND	0.02 ng/L	
					d Kouchibouguac, Prince	DDT, p,p'-	a, b, d	0.02 ng/L	
					Edward Island, Canada	Dieldrin	All	0.02 ng/L	
						Endosulfan (I+II)	ND	0.02 ng/L	
						Endrin	All	0.02 ng/L	
						HCB	All	0.02 ng/L	
						HCH, α-	All	0.02 ng/L	
						HCH, γ-	All	0.02 ng/L	
						Heptachlor epoxide	All	0.02 ng/L	
						Methoxychlor	All	0.02 ng/L	
						Mirex	ND	0.02 ng/L	
						Toxaphene	ND	2.0 ng/L	

Table 2.4. Characteristics and summaries of national and multistate pesticide monitoring studies—*Continued*

Study no.	Study	Sampling Matrix	Sampling Date	Sampling Quantity	Location(s)	Compounds Name	Compounds Occurrence	Detection limits	Comments
9	Strachan, 1990	Rain Snow	1983-86 (17 sites around the Great Lakes and throughout Canada)	NR	a Kanala Creek, British Columbia, Canada b Suffield, Alberta, Canada c Cree Lake, Saskatchewan, Canada d Isle Royale, Lake Superior e Caribou Island, Lake Superior f Kouchibouguac, National Park, New Brunswick, Canada	Aldrin Chlordane, *cis-* Chlordane, *trans-* DDD, *p,p'-* DDE, *p,p'-* DDT, *p,p'-* Dieldrin Endosulfan I Endosulfan II Endrin HCB HCH, α- HCH, γ- Heptachlor Heptachlor epoxide Methoxychlor Mirex	ND ND ND All but f ND All ND ND ND ND ND All but f All All ND ND ND	For all compounds: 0.02 ng/L	Reported values for α-HCH, γ-HCH, dieldrin, *p,p'-*DDE, HCB, PCBs. Found that the relative concentration for all compounds were similar at all of the sites throughout 1983-86. Occurrence of the other compounds was sporadic at most of the sites.
10	Sweet, 1992	Air Rain Snow Dry-Dep.	Oct 1990 to present	439	a Point Petre, Lake Ontario, Canada b Eagle Harbor, Lake Superior, MI c Sturgeon Pt., Lake Erie, NY d Sleeping Bear, MI	DDD, *p,p'-* DDE, *p,p'-* DDT, *p,p'-* Dieldrin HCB HCH, α- HCH, γ-	All All All All All All All	ND NR <0.1 ng/L (rain) <1 pg/m^3 (air) <1 mg/m^3 (ptcl.) (all inferred from data)	A preliminary report with data presented at an international meeting. Primary information was on PCB depositions into the Great Lakes.

Table 2.4. Characteristics and summaries of national and multistate pesticide monitoring studies--*Continued*

Study no.	Study	Sampling			Location(s)	Compounds		Detection limits	Comments
		Matrix	Date	Quan-tity		Name	Occurrence		
11	Goolsby and others, 1994	Rain Snow	Mar 1990-Sept 1991 (1 week cumu-lative samples)	6,100	81 National Atmospheric Deposition Program sampling sites in 24 states in the northeastern United States	Alachlor	19*	For all compounds:	Investigated the local/regional transport and distribution of corn belt herbicides.
						Ametryn	ND		
						Atrazine	30*	0.05 µg/L	
						Cyanazine	6.2*		
						Desethyl-atrazine	6.2*		
						Desisopropyl-atrazine	2.6*		
						Metolachlor	13*		
						Metribuzin	0.6*		
						Prometon	0.4*		
						Prometryn	ND		
						Propazine	0.1*		
						Simazine	1.2*		
						Terbutryn	ND		
						(*, percent detections)			

Table 2.4. Characteristics and summaries of national and multistate pesticide monitoring studies--*Continued*

Study no.	Study	Sampling				Compounds			Comments
		Matrix	Date	Quan-tity	Location(s)	Name	Occurrence	Detection limits	
12	Yeary and Leonard, 1993	Air	No dates given	500	Indoor air of 82 homes; outdoor air of 55 homes; breathing zone air of 200 pesticide applicators; 49 pesticide office/warehouse facilities. a Baltimore, MD b Buffalo, NY c Chicago, IL d Columbus, OH e Dallas, TX f Fort Wayne, IN g Los Angeles, CA h Monmouth, NJ i Montreal, Quebec, Canada j Orlando, FL k Philadelphia, PA l Tampa, FL m Toronto, Ontario, Canada n Washington, D.C.	Acephate Atrazine Ammonia Bensulide Carbaryl Chlorpyrifos 2,4-D Dacthal Diazinon Dicofol MCPA Pendimethalin	2 of 34 sites 7 of 22 sites 0 of 12 sites 0 of 10 sites 29 of 66 sites 9 of 39 sites 30 of 143 sites 0 of 2 sites 18 of 100 sites 2 of 116 sites 0 of 37 sites 1 of 10 sites	For all compounds: 0.001 mg/m^3	Investigated airborne levels of selected pesticides and formulation ingredients in the breathing zones of pesticide applicators, indoor air of pesticide warehouse facilities and offices, and indoor and ambient air of residential properties.

CHAPTER 3

National Distributions and Trends

The studies reviewed show that one or more classes of pesticides have been found in every region of the nation where they were investigated (Tables 2.2, 2.3, and 2.4). Although no study analyzed for every pesticide, or even representative pesticides from every class, the studies, taken together, show that a wide variety of pesticides are present in air, rain, fog, and snow. The process and matrix distribution studies (Table 2.2), as described in the previous section, do not provide much information on occurrence and distribution of pesticides in the atmosphere because they generally tested new sampling devices or sampled a particular field application. The state and local monitoring studies (Table 2.3) provide most of the available information on occurrence and distribution of pesticides in the atmosphere, but the results of these studies are difficult to compare because of the differing study attributes such as sampling matrix, sites, times, target analytes, and analytical methods.

3.1 PESTICIDES DETECTED

The best method of assessing occurrence and distribution on a national scale is through national and multistate monitoring studies (Table 2.4), but only five have been done in the United States since the 1960's and none of these covered the entire nation. Tabor (1965) sampled air in eight agricultural communities and in four urban communities with active insect control programs. He detected a wide variety of pesticides, both organochlorine and organophosphorus insecticides, in the air where they were expected because of the proximity of the community to high agricultural use or local use. Stanley and others (1971) sampled air at five rural and four urban sites in eight states for 16 organochlorine compounds, 4 organophosphorus compounds, and several 2,4-D esters during 1967. The sampling was designed to provide information on the scope and type of pesticide contamination in the air. DDT was detected in all samples (2.7 ng/m^3 to 1,560 ng/m^3), and heptachlor epoxide, chlordane, DDD, and 2,4-D esters were not detected in any sample. The four organophosphorus compounds (parathion, methyl parathion, diazinon, and DEF) were detected only at rural sites in three southern states. Parathion and methyl parathion were detected most frequently and at the highest concentrations. DDT and several of its transformation products, HCH isomers (α-, β-, γ- and δ-), and toxaphene also were detected at the same three sites. Organochlorine insecticides also were detected in the two northern agricultural sites, but at concentrations approximately one to two orders of magnitude less than in the southern states. DDT's and HCH's were detected in four urban areas at concentrations comparable to those detected in northern rural sites. The highest pesticide concentrations in air among nine cities in eight states were found in the agricultural areas of the south, with generally lower concentrations in urban areas. They expected the concentrations in Fresno and Riverside, California, to be higher than measured but their sampling sites apparently were too far away from the intense agricultural activity in these areas. The findings in this study indicate that pesticide air concentrations generally were highest when spraying was reported to occur during the same period as air sampling. Concentrations were highest during the summer months at all locations.

Kutz and others (1976) sampled air at 28 sites in 16 states between 1970 and 1972. The sampling locations included urban and rural sites, but each site was selected for its potential for high pesticide concentrations in ambient air. Samples were analyzed for 18 organochlorines; 7 organophosphates; 3 esters of 2,4-D; 2 esters of 2,4,5-T; dacthal; and trifluralin. Not all sampling sites were used during the 3-year study, and the number of pesticides detected and their mean concentrations varied greatly between states. DDT, several of its transformation products, dieldrin, and α-HCH were detected in nearly every sample. All of these compounds except α-HCH were present at similar concentrations throughout the year. Aldrin, dieldrin, γ-HCH, and heptachlor were detected in every state, but not at every site, as were diazinon, malathion, methyl parathion, and trifluralin. Toxaphene was detected only in the southern states, but it was found in the highest concentrations of any detected pesticide.

Richards and others (1987) collected rain samples from four locations in four states during the spring and summer of 1985. These samples were analyzed for 19 compounds that represented 90 percent (by weight) of the pesticides used in Ohio. Eleven compounds were detected: four triazines, one organophosphate, five other herbicides, and one other insecticide. The Indiana and Ohio sites were in intensive agricultural areas, whereas the West Virginia and New York sites were in less intensive agricultural areas. Alachlor, atrazine, cyanazine, and simazine were detected at all sites, but the concentrations of each herbicide were highest in Indiana and Ohio. The overall findings for each of the nine herbicides and two insecticides detected in precipitation were that the samples from West Virginia and New York had lower concentrations and less frequent detections than those collected in Indiana and Ohio.

The latest reported multistate study was done in 1990 and 1991 by Goolsby and others (1994). They sampled at 81 National Atmospheric Deposition Program/National Trends Network (NADP/NTN) sites in 23 states in the north-central and northeastern United States. Their analyses, however, were restricted to those herbicides used in corn and soybean production: the triazines and acetanilides. Their results showed that the highest triazine concentrations occurred in six major corn-producing states: Minnesota, Nebraska, Wisconsin, Illinois, Iowa, and Indiana, with average precipitation-weighted concentrations ranging from 100 to 500 ng/L during spring and early summer. The maximum concentration was as high as 3,000 ng/L. Concentrations in Nebraska and northern Illinois precipitation were greater than 500 ng/L during this same time period. Average precipitation-weighted acetanilide concentrations also were detected in the same six-state area and ranged from 100 to 700 ng/L. The highest concentrations, ranging from 1,000 to 3,000 ng/L, occurred in northern Illinois, northeastern Iowa, and northeastern Indiana.

In order to examine general occurrence and distribution patterns for pesticides in the atmosphere, results from all state and local monitoring studies were combined with data from the national and multistate occurrence studies. Figure 3.1 shows those pesticides that have been analyzed for in air and rain at 10 or more sites in the United States, and the percentage of sites at which they were detected. The number and variety of pesticides detected in air (5 herbicides, 20 insecticides, 1 fungicide, 7 metabolites) are greater than that detected in rain (6 herbicides, 5 insecticides, 0 fungicides, 3 metabolites). This disparity does not suggest that rain contains less pesticides than air, because an equal number of pesticides have been detected in rain (Tables 2.2,

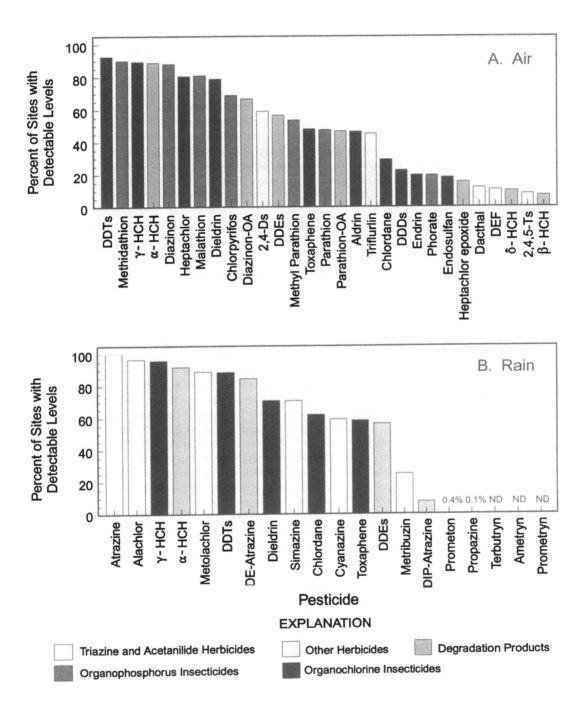

FIGURE 3.1. National occurrence and detection frequency of pesticides analyzed for in (A) air and (B) rain at 10 or more sites.

2.3, and 2.4). It indicates that rain has not been analyzed for pesticides as often and at as many sites as air has been. Also, less rain water is extracted for analysis than for air, so the analytical limits of detection are generally higher.

Figure 3.2 shows the number of different pesticides detected in air, rain, snow, and fog per state from each of the four major pesticide categories from available data. It also shows that most studies on pesticide occurrence in the atmosphere have been done in the eastern half of the United States. The greatest number of studies that analyzed for organophosphorus insecticides and the greatest variety of organophosphorus insecticides detected occurred in California. Nationwide, the organochlorine insecticides and herbicides were the predominant pesticides analyzed for and detected. This does not mean that the organophosphorus insecticides and other pesticides are not present in the atmosphere, or that no pesticides were present in those states with no data. Organophosphorus insecticides have not been analyzed for as frequently as the other classes. As described for Figure 3.1, the distribution of results is highly affected by the distribution of study efforts. The occurrence of few or no detections in a state does not necessarily imply absence.

3.2 SUMMARY OF NATIONAL USE

Table 3.1 lists the pesticides that were used in agriculture in quantities of 8,000 or more lb a.i. in 1988 (Gianessi and Puffer, 1990, 1992a,b) along with United States use estimates for 1971 and 1966 (Andrilenas, 1974). Those pesticides that have been detected in rain and air are indicated. One-hundred pesticides (57 herbicides, 31 insecticides, 12 fungicides) were used in quantities of greater than 500,000 lb a.i. in 1988. Of these, only 19 were analyzed for at 10 or more sites and 17 were detected at least once. Three insecticides with use of less than 500,000 lb a.i. in 1988 also were detected in air, and two of the three were detected in rain at 10 or more sites. Of more than 150 pesticides listed in Table 3.1, only 47 have been analyzed for and of these, 40 have been detected. In all of the studies listed in Tables 2.2, 2.3, and 2.4, excluding the field worker exposure, volatility, and drift studies where a known amount of a specific compound was applied to a specific area, 63 pesticides have been analyzed for and 45 have been detected. Eighteen pesticide transformation products were also detected.

Table 3.1 includes the number of outdoor applications of pesticide active ingredients around homes and gardens for 1990 (Whitmore and others, 1992). The study from which these data were compiled was not designed to collect quantitative pesticide use information, but only to inquire as to the products each interviewed household had on hand and the number of times they used each of them. An important point that emerges from a comparison of the top 50 pesticides used in agriculture versus the home and garden is that there is only about a 20 percent similarity in the types of pesticides used. The main types of pesticides used in agriculture differ considerably from those used in and around the home and garden. This is partly due to U.S. Environmental Protection Agency (USEPA) use restrictions.

3.3 GEOGRAPHIC DISTRIBUTION IN RELATION TO USE

Occurrence and geographic distribution of pesticides in air and rain, and the relation between the measured air concentrations, detection frequencies, and national use can best be examined by comparing results from individual national and multistate occurrence studies to national pesticide use patterns. Only two studies, however, present data in such a way that this

FIGURE 3.2. Number of different pesticides detected in air, rain, snow, and fog per state by major class: (A) organochlorine insecticides, (B) organophosphorus insecticides, (C) triazine and acetanilide herbicides, and (D) other herbicides.

TABLE 3.1. Agricultural pesticides used in the United States, in thousands of pounds of active ingredient (a.i.) for 1966, 1971, and 1988 (Gianessi and Puffer, 1990, 1992a,b), and urban pesticides used in and around the home and garden (Whitmore and others, 1992) in 1990 in thousands of products used and number of outdoor applications

[Agricultural use is ranked in descending order of use for 1988 except for organochlorine insecticides which are ranked in descending order of use for 1971. Those pesticides detected in air or rain from studies listed in Tables 2.2, 2.3, and 2.4 are noted. Blank spaces indicate no data]

Compound	Detected in		Agricultural use			1990 Home use	
			Pounds active ingredient (× 1,000)			Products (× 1,000)	Outdoor applications (× 1,000)
	Rain	Air	1966	1971	1988		
ORGANOCHLORINE INSECTICIDES							
Toxaphene	R	A	34,605	37,464		74	
DDT	R	A	27,004	14,324		202	
Aldrin	R	A	14,761	7,928			
Methoxychlor	R	A	2,578	3,012	109	3,564	3,692
Chlordane	R	A	526	1,890		1,156	478
Endrin		A	571	1,427			
Heptachlor		A	1,536	1,211		72	177
Endosulfan	R	A	791	882	1,992	111	561
HCH, γ-	R	A	704	650	66	1,638	1,355
Dieldrin	R	A	724	332			
DDD		A	2,896	244			
Strobane			2,016	216			
Others			347	293			
Dicofol					1,718	4,587	4,179
Paradichlorobenzene						1,098	538
Pentachlorophenol (total)						576	89
Ronnel						38	
Total use			89,059	69,873	3,885	13,116	11,069
Percent of total insecticide use			63	43	4	4	2
ORGANOPHOSPHORUS INSECTICIDES							
Chlorpyrifos	R	A			16,725	16,652	41,900
Methyl Parathion	R	A	8,002	27,563	8,131		
Terbufos					7,218		
Phorate		A	326	4,178	4,782		
Fonofos	R	A			4,039		
Malathion	R	A	5,218	3,602	3,188	9,551	16,597
Disulfoton		A	1,952	4,079	3,058	2,364	6,464
Acephate					2,965	4,940	19,167
Dimethoate	R				2,960	301	132
Parathion	R	A	8,452	9,481	2,848		
Azinphos-methyl			1,474	2,654	2,477	37	348
Diazinon	R	A	5,605	3,167	1,710	15,703	56,758
Ethoprop					1,636		
Ethion			2,007	2,326	1,249	39	
Profenfos					1,224		
Methamidophos					1,135		
Phosmet					1,055	371	173
Dicrotophos			1,857	807	963		

TABLE 3.1. Agricultural pesticides used in the United States, in thousands of pounds of active ingredient (a.i.) for 1966, 1971, and 1988 (Gianessi and Puffer, 1990, 1992a,b), and urban pesticides used in and around the home and garden (Whitmore and others, 1992) in 1990 in thousands of products used and number of outdoor applications--*Continued*

Compound	Detected in		Agricultural use			1990 Home use	
			Pounds active ingredient (× 1,000)			Products (× 1,000)	Outdoor applications (× 1,000)
	Rain	Air	1966	1971	1988		
ORGANOPHOSPHORUS INSECTICIDES--*Continued*							
Sulprofos					874		
Fenamiphos					763		
Mevinphos					463		
Methidathion		A			402		
Oxydemeton-methyl					370	1,032	670
Naled					224	158	
Trichlorfon			1,060	617	16	41	
Dichlorvos			912	2,434		8,953	13,043
Ronnel			391	479			
Others			2,710	9,319			
Tetrachlorvinphos						2,423	
Phosalone						118	93
Chlorfenvinphos						105	
Isofenphos						100	43
Crotoxyphos						40	
Total use:			39,966	70,706	70,475	62,928	155,388
Percent of total insecticide use:			28	44	65	19	26
OTHER INSECTICIDE AND NONHERBICIDE PESTICIDES							
Carbaryl		A	12,392	17,838	7,622	18,437	31,735
Carbofuran	R			2,854	7,057		
Propargite					3,786		
Aldicarb					3,573		
Cryolite					2970		
Methomyl				1,077	2,952	346	629
Thiodicarb					1,714		
Permethrin					1,122	7,397	18,461
Oxamyl					726		
Fenbutatin oxide					560	248	2,147
Formethanate HCl					414		
Esfenvalerate					287		
Tefluthrin					197		
Cypermethrin					188	1,98	
Trimethacarb					131		
Lambdacyhalothrin					110		
Cyfluthrin					105	1,139	5,654
Oxythioquinox					102		
Amitraz					75		
Fenvalerate					73	3,192	2,937
Metaldehyde					44	5,144	27,094
Tralomethrin					43		
Diflubenzuron					42		

TABLE 3.1. Agricultural pesticides used in the United States, in thousands of pounds of active ingredient (a.i.) for 1966, 1971, and 1988 (Gianessi and Puffer, 1990, 1992a,b), and urban pesticides used in and around the home and garden (Whitmore and others, 1992) in 1990 in thousands of products used and number of outdoor applications--*Continued*

| Compound | Detected in | | Agricultural use | | | 1990 Home use | |
| | Rain | Air | Pounds active ingredient (× 1,000) | | | Products (× 1,000) | Outdoor applications (× 1,000) |
			1966	1971	1988		
OTHER INSECTICIDE AND NONHERBICIDE PESTICIDES--*Continued*							
Bifenthrin					28		
Abamectin					12		
Cyromazine					9		
Others			502	37			
Piperonyl butoxide						41,729	58,991
Pyrethrins						34,609	39,289
MGK-264						27,558	13,249
Diethyltoluamide						21,544	14,134
Propoxur						21,484	53,594
Allethrin, (total)						18,543	52,277
Tetramethrin						12,962	31,464
Resmethrin, (total)						12,506	34,576
Sumithrin						8,089	3,1856
Rotenone						3,997	4,510
Methoprene						2,709	1,999
Hydramethylon						2,389	10,485
Allethrin						1,388	2,005
Warfarin						1,145	545
Brodifacoum						906	296
Bendiocarb						697	1,778
Tricosene						307	453
Dienochlor						265	1,591
Methiocarb						237	386
Diphacinone						89	
Pinadone						81	
Bromadiolone						70	177
Fenoxycarb						54	
Cythioate						37	
Total use:			12,894	21,806	33,941	249,496	442,312
Percent of total insecticide use:			9	13	31	77	73
TRIAZINE AND ACETANILIDE HERBICIDES							
Atrazine	R	A	23,521	57,445	64,236	134	488
Alachlor	R	A		14,754	55,187		
Metolachlor	R	A			49,713		
Cyanazine	R				22,894		
Metribuzin	R				7,516		
Propazine	R		580	3,171	4,015		
Propachlor	R		2,269	23,732	3,989		
Simazine	R	A	193	1,738	3,964	172	
Prometryn					1,807		

TABLE 3.1. Agricultural pesticides used in the United States, in thousands of pounds of active ingredient (a.i.) for 1966, 1971, and 1988 (Gianessi and Puffer, 1990, 1992a,b), and urban pesticides used in and around the home and garden (Whitmore and others, 1992) in 1990 in thousands of products used and number of outdoor applications--*Continued*

Compound	Detected in		Agricultural use			1990 Home use	
			Pounds active ingredient (× 1,000)			Products (× 1,000)	Outdoor applications (× 1,000)
	Rain	Air	1966	1971	1988		
TRIAZINE AND ACETANILIDE HERBICIDES--*Continued*							
Terbutryn					1,113		
Ametryn					186		
Prometon	R					1,244	1,281
Other				450	1,022		
Total use:			26,563	101,290	215,642	1,550	1,769
Percent of total herbicide use:			24	44	47	3	1
OTHER HERBICIDES							
EPTC	R		3,138	4,409	37,191	125	
2,4-D		A	40,144	34,612	33,096	14,324	44,054
Trifluralin	R	A	5,233	11,427	27,119	483	547
Butylate	R			5,915	19,107		
Pendimethalin	R	A			12,521	158	338
Glyphosate					11,595	8,110	25,618
Dicamba (Canada)	R		222	430	11,240	3,636	6,431
Bentazon					8,211		
Propanil			2,589	6,656	7,516		
MSMA					5,065	240	267
Molinate		A			4,408		
MCPA			1,669	3,299	4,338		
Ethafluralin					3,518		
Triallate (Canada)		A			3,509		
Paraquat					3,025	201	79
Chloramben			3,765	9,555	3,019		
Picloram					2,932		
Clomazone					2,715		
Bromoxynil (Canada)	R				2,627		
Linuron			1,425	1,803	2,623		
Fluometuron				3,334	2,442		
Dacthal	R	A			2,219	733	368
Diuron			1,624	1,234	1,986		
Norflurazon					1,768		
DSMA					1,705	38	38
Acifluorfen					1,475	1,845	7,081
Diclofop (Canada)	R				1,452		
Oryzalin					1,426	117	1,766
2,4-DB					1,368		
Thiobencarb		A			1,359		
Cycloate					1,175		
Benefin					1,167	79	81
Bromacil					1,155		
Asulam					1,088		
Imazaquin					1,073		

TABLE 3.1. Agricultural pesticides used in the United States, in thousands of pounds of active ingredient (a.i.) for 1966, 1971, and 1988 (Gianessi and Puffer, 1990, 1992a,b), and urban pesticides used in and around the home and garden (Whitmore and others, 1992) in 1990 in thousands of products used and number of outdoor applications--*Continued*

Compound	Detected in		Agricultural use			1990 Home use	
			Pounds active ingredient (× 1,000)			Products (× 1,000)	Outdoor applications (× 1,000)
	Rain	Air	1966	1971	1988		
OTHER HERBICIDES--*Continued*							
Diphenamid					929		
Vernolate				3,739	855		
Sethoxydim					792		
Fluazifop					731		
Napropamide					699		
Naptalam			999	3,332	655		
Pebulate			150	1,062	653		
Bensulide					633	36	73
Profluralin					621		
Tebuthiuron					608		
Oxyfluorfen					599	234	643
Diethatyl ethyl					502		
Dalapon			38	1,043	453		
Alanap			999	3,332			
Nitralin			14	2,706			
2,4,5-T	R	A	760	1,530		84	
Fluorodifen				1,330			
Norea			239	1,323			
Other			21,479	27,686	5,635		
MCPP (total)						11,843	32,378
Mecoprop						849	3,266
Triclopyr, (total)						683	823
Chlorflurenol, methyl ester						501	1,067
Fluazifop-butyl						421	426
Silvex, (total)						341	88
Diquat dibromide						211	635
Dichlobenil						167	124
Amitrole						163	40
Metam-Sodium						128	74
Endothal, di-Na salt						124	43
MCPA, dimethylamine salt						119	121
Sodium thiocyanate						82	
Total use:			84,487	129,757	242,598	46,075	126,469
Percent of total herbicide use:			76	56	53	97	99
FUNGICIDES							
Chlorthalonil					9,932	1,399	2,602
Mancozeb					8,661	113	
Captan			6,869	6,490	3,710	3,067	4,682
Maneb			4,443	3,878	3,592	345	878
Ziram					1,889	81	
Benomyl					1,344	684	3,704

TABLE 3.1. Agricultural pesticides used in the United States, in thousands of pounds of active ingredient (a.i.) for 1966, 1971, and 1988 (Gianessi and Puffer, 1990, 1992a,b), and urban pesticides used in and around the home and garden (Whitmore and others, 1992) in 1990 in thousands of products used and number of outdoor applications--*Continued*

Compound	Detected in		Agricultural use Pounds active ingredient (× 1,000)			1990 Home use	
	Rain	Air	1966	1971	1988	Products (× 1,000)	Outdoor applications (× 1,000)
FUNGICIDES--*Continued*							
PCNB					800	64	64
Iprodione					741		
Fosetyl-Al					689		
Metiram					641		
Metalaxyl					635	41	
Thiophanate methyl					527		
Triphenyltin hydroxide					415		
Ferbam			2,945	1,398	337	82	
DCNA					286		
Dodine **			1,143	1,191	275		
Propiconazole					274		
Thiram					238	397	249
Triadimefon					149		
Anilazine					144		
Thiabendazole					139		
Myclobutanil					124		
Etridiazole					104		
Vinclozolin					103		
Streptomycin					88		
Triforine					81	3,150	4,347
Fenarimol					58		
Oxytetracycline					37		
Carboxin					15		
Dinocap **			1,143	1,191	14		
Zineb			6,903	1,969		684	3,704
Other			3,334	10,814			
Folpet						3,314	4,347
Limonene						819	
Dinocap						703	796
Anilazine						106	71
Dextrin						79	
Dichlorophene						43	
Dichlone						40	
Thymol						37	166
Total fungicide use:			26,780	26,931	36,042	15,248	25,610
(**Use for these two compounds was reported as combined totals for 1966 and 1971.)							

can be done. The geographic distribution and use relation of the organochlorine and organophosphorus insecticides, and other herbicides, can best be examined using the air concentration data of Kutz and others (1976) for 1970-72 and pesticide use data for 1971 from the U.S. Department of Agriculture (Andrilenas, 1974). Those organochlorine insecticides that were detected at a frequency of 50 percent or more and those organophosphorus insecticides and other herbicides that were detected at a frequency of 10 percent or more are discussed in detail. The geographical distribution and use relation of triazine and acetanilide herbicides can best be examined using the 1990-91 rain concentration data from Goolsby and others (1994) and the 1988 herbicide use data from Gianessi and Puffer (1990).

Because the 1971 pesticide use data were compiled by agricultural region (Table 3.2) instead of by state or county, it is difficult to make any accurate assessments. Only broad, general patterns become apparent when air concentrations, detection frequencies, regional use, and major regional cropping patterns are compared. The 1988 data is more refined, in that the sampling coverage was more extensive and the herbicide use data were much more specific.

ORGANOCHLORINE INSECTICIDES

Because of their widespread use during the 1960's and 1970's, and their resistance to environmental transformation, organochlorine insecticides have been detected in the atmosphere of every state in which they have been looked for (Figure 3.2A). The most heavily used organochlorine compounds during this time period were toxaphene, DDT, and aldrin (Table 3.1), but their total use in agriculture has declined steadily from 63 percent of the total insecticides used in 1966 to less than 5 percent in 1988 (Table 3.1) because of reduced effectiveness or regulatory restrictions, or both. Figure 3.3 shows the frequency of detections in air and rain in relation to total use for those organochlorine insecticides detected at 10 or more sites. The

TABLE 3.2. U.S. Department of Agriculture farm production regions as defined for 1971 agricultural pesticide use (Andrilenas, 1974)

Agricultural region	State
Appalachian	Kentucky, North Carolina, Tennessee, Virginia, West Virginia
Corn belt	Illinois, Indiana, Iowa, Missouri, Ohio
Delta states	Arkansas, Louisiana, Mississippi
Lake states	Michigan, Minnesota, Wisconsin
Mountain	Arizona, Colorado, Idaho, Montana, Nevada, New Mexico, Utah, Wyoming
Northeast	Connecticut, Delaware, Maine, Maryland, Massachusetts, New Hampshire, New Jersey, New York, Pennsylvania, Rhode Island, Vermont
Northern plains	Kansas, Nebraska, North Dakota, South Dakota
Pacific	California, Oregon, Washington
Southeast	Alabama, Florida, Georgia, South Carolina
Southern plains	Oklahoma, Texas

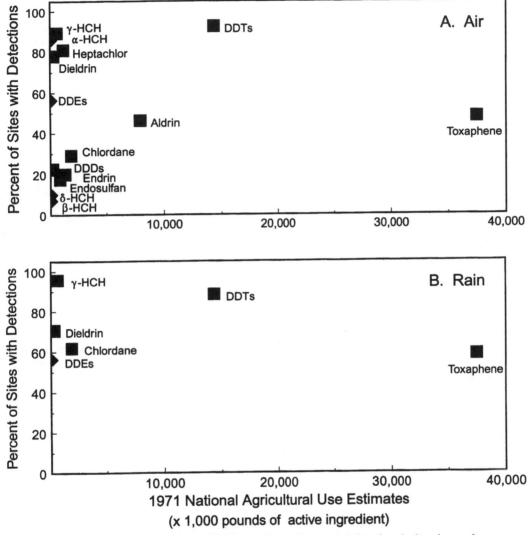

FIGURE 3.3. The relation between site detection frequency and national agricultural use for organo-chlorine insecticides and degradation products detected in (A) air and (B) rain. The detection frequency data are for those compounds analyzed at 10 or more sites, and the use data are from the 1971 U.S. Department of Agriculture-National Agricultural Statistics Service agricultural pesticide use estimates (Andrilenas, 1974). ■ denotes pesticide active ingredient and ◆ denotes pesticide degradation product.

frequency of detection refers to the percent of sites with detections, and not to the percent of samples with detections. The most frequently detected organochlorine insecticides have been DDT, α-HCH, γ-HCH, heptachlor, and dieldrin. Figure 3.3A shows that both toxaphene and aldrin were detected at less than 50 percent of the sites where they were sampled for. Why these insecticides were detected less frequently than DDT can be explained partially by the chemistry of these compounds. Toxaphene is a complex mixture of over 200 different compounds, and obtaining a representative sample and maintaining its integrity throughout sampling and analysis is an involved and difficult process. Because toxaphene was banned in 1982 and has a complex nature, the analytical "fingerprint" of environmental samples is often considerably different from the analytical standards due to physical weathering, chemical degradation, and metabolism (Bidleman and others, 1988). Also, the analytical limit of detection for toxaphene is considerably higher than for single component organochlorine insecticides.

The sampling and analysis of DDT and its metabolites DDD and DDE, on the other hand, are relatively straightforward. This may be a reason why 37 of 44 studies (not including the field volatility experiments) listed in Tables 2.2, 2.3, and 2.4, that analyzed for organochlorine compounds, included DDT's as target analytes while only 17 included toxaphene. Toxaphene's concentrations in air (0.01 to 1,540 ng/m^3) and rain (less than 2 to 497 ng/L) spanned a wide range, with the highest levels measured in the cotton-growing regions of the southern United States during the summer months.

Aldrin detections were only reported in four studies that were done during the late 1960's (Tabor, 1965; Stanley and others, 1971) and early 1970's (Arthur and others, 1976; Kutz and others, 1976). Tabor (1965) detected aldrin at concentrations of 1 ng/m^3 in 50 percent of the air samples taken at a city in Florida centered in a winter vegetable-producing area and surrounded by vegetable fields. Aldrin was not detected at five other agricultural communities where air sampling was also done. Stanley and others (1971) detected aldrin in only one air sample (8.0 ng/m^3), taken at a rural site in Iowa, out of 880 samples from nine states. Kutz and others (1976) detected aldrin in 15 out of 16 different states (0.8 to 28 ng/m^3) in which sampling was done, but the sites were chosen for their potential of having high pesticide concentrations in the ambient air. They found that the mean concentration levels and the detection frequency declined during the study. Arthur and others (1976) reported a maximum aldrin air concentration of 6.9 ng/m^3 but did not discuss how often it was detected during a 3-year study at Stoneville, Mississippi. Shulters and others (1987) analyzed for aldrin in precipitation at Fresno, California, but none was detected. The relatively low frequency of aldrin detections in air and precipitation may be due to its conversion to dieldrin and other products in the atmosphere (Crosby and Moilanen, 1974). Dieldrin, which had relatively low direct use, has been reported more frequently (Figure 3.1A) at similar concentration ranges (0.7 to 29.7 ng/m^3) as aldrin for the same time period, 1967-74. It has been more frequently included in the list of analytes measured in air, rain, and snow samples in Canadian studies (Strachan and Huneault, 1979; Chan and Perkins, 1989; Strachan, 1990; Brun and others, 1991) and although its detection frequencies were high (80 percent), the concentration levels usually were low.

The HCH isomers (α-, β-, γ- and δ-) are another group of organochlorine compounds that are frequently detected in rain and air, especially relative to their use. Technical lindane, which includes these four HCH isomers and others, is no longer used in the United States and Canada. It has been replaced with lindane, which is greater than 99.5 percent pure γ-HCH. The major atmospheric conversion product of γ-HCH is the α-HCH isomer, and both of these are often analyzed for together and have been detected throughout the United States and Canada in air (α- 0.049 to 11.0 ng/m^3; γ- 0.006 to 7.0 ng/m^3) and precipitation (α- 1 to 88 ng/L; γ- 0.3 to 40 ng/L). The use of γ-HCH has never been very high (Table 3.1), but it has been fairly constant, whereas the use of other organochlorine pesticides, such as DDT, aldrin, and toxaphene, has been severely restricted or banned. Also, γ-HCH's vapor pressure and water solubility are higher than many other organochlorine pesticides (Suntio and others, 1988); therefore, the applied residues will volatilize more rapidly and be redeposited by rain and snow. The vapor pressure of heptachlor is also high relative to the other compounds shown in Figure 3.3A, which may help explain the high detection frequency in air relative to its use.

The overall conclusion from Figure 3.3 is that there does not appear to be any straightforward relation between the use of an organochlorine insecticide and its detection in air or rain. The main reason for this appears to be inconsistency in the spatial and temporal coverage of the national studies. Many of the inconsistencies could be explained, however, by the relation between a parent compound and one or more metabolites such as DDT and DDE, γ-HCH and α-HCH, and aldrin and dieldrin. Sampling and analytical difficulties and high detection limits also help explain some discrepancies, such as the case of toxaphene.

When the data of Kutz and others (1976) are examined more closely, a better detection-to-use relation is evident for those insecticides with detection frequencies of 50 percent or more. Figure 3.4 shows the relation of p,p'-DDT, o,p'-DDT, and p,p'-DDE detection frequencies (98, 82, and 96 percent, respectively) to 1971 use by agricultural production region (Table 3.2). The agricultural regions defined in this table will be used in the discussions throughout the remainder of this chapter. Technical DDT contained up to 30 percent o,p'-DDT, which has similar insecticidal characteristics as p,p'-DDT. p,p'-DDE is a major metabolite of p,p'-DDT. DDT was primarily used on cotton in the Southeast and Delta states. It was also used in the Appalachian and Southern Plain states on soybeans and grains and, to a lesser extent, on wheat, tobacco, and corn. DDT's regional use corresponds with a majority of the high averaged air concentrations at sites located in the Delta, Southeast, and Appalachian regions. o,p'-DDT and p,p'-DDE follow the same air concentration distribution as p,p'-DDT at near equivalent levels. The air concentration of o,p'-DDT is nearly one-third that of p,p'-DDT, which corresponds to its percentage in technical DDT. The detection frequencies of p,p'-DDE are nearly equivalent to p,p'-DDT and were routinely above 95 percent at most sites, but o,p'-DDT detection frequencies were nearly 20 percent less.

Aldrin and dieldrin are two insecticides detected by Kutz and others (1976) with detection frequencies of 16 and 91 percent, respectively. The detection frequency of aldrin is low relative to its use (Figure 3.5A), but its chemistry and detection are closely related to dieldrin, as previously discussed. The average air concentration and detection frequency per site for both insecticides are shown in Figure 3.5. Dieldrin values (Figure 3.5B) are shown with aldrin use data because of the relative insignificance of dieldrin's actual use and because dieldrin is a major breakdown product of aldrin.

Aldrin primarily was used on corn, with much less used on soybeans and tobacco. Cotton was the primary crop that dieldrin was used on. Aldrin air concentrations are not adequately explained by its regional use and cropping patterns. It was detected at high concentrations throughout the nation, in both high and low use areas. The detection frequencies were usually low, however, and there were many sites where it was not detected at all. Some of the sites where no aldrin was detected were near others with high air concentrations (Figure 3.5A), however, and the reasons for this are unclear. Although the sampling sites were located in agricultural areas, not all pesticides on the analyte list were necessarily used near each site. More detailed cropping patterns and pesticide use numbers would help clarify this. Kutz and others (1976) observed that aldrin air concentrations and detection frequencies declined during the 3-year study while dieldrin concentrations and detection frequencies remained constant. This could reflect the decline in aldrin's use during this period. At every site where no aldrin was detected, however, dieldrin was detected in greater than 85 percent of the samples. Dieldrin also was detected at nearly every site and at very high detection frequencies. High dieldrin air concentrations occurred in the lowest dieldrin use areas (the Northern Plains, the Lake States, and the Corn Belt regions), but these regions corresponded to high aldrin use and high aldrin air concentrations. The high detection frequencies and air concentrations reflect dieldrin's longevity in the environment as well as the fact that it is a major conversion product of aldrin.

γ-HCH and α-HCH are another pair of organochlorine insecticides that Kutz and others (1976) detected at frequencies of 71 and 91 percent, respectively, and whose detection frequencies and concentrations are interrelated. γ-HCH was primarily used on fruits and nuts in the Delta and Southeast regions. These areas correspond to the highest average air concentrations and the highest detection frequencies (Figures 3.6A and 3.6B). γ-HCH was also used on livestock

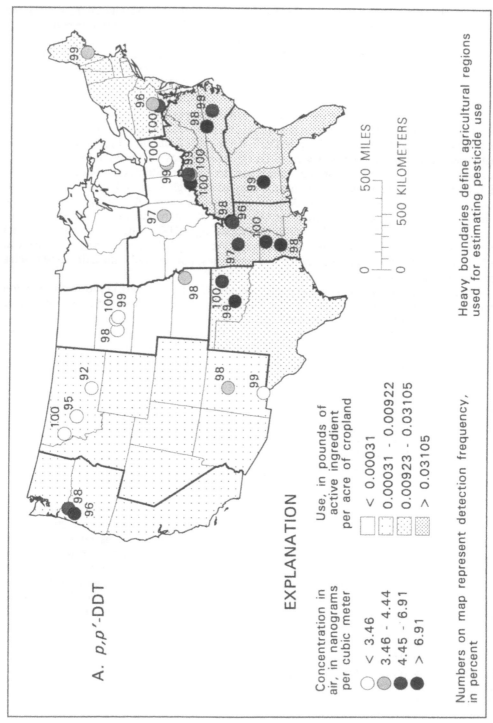

FIGURE 3.4. Average range of measured concentrations of (A) *p,p'*-DDT, (B) *o,p'*-DDT, and (C) *p,p'*-DDE in air and the detection frequency at each sampling site of Kutz and others (1976).

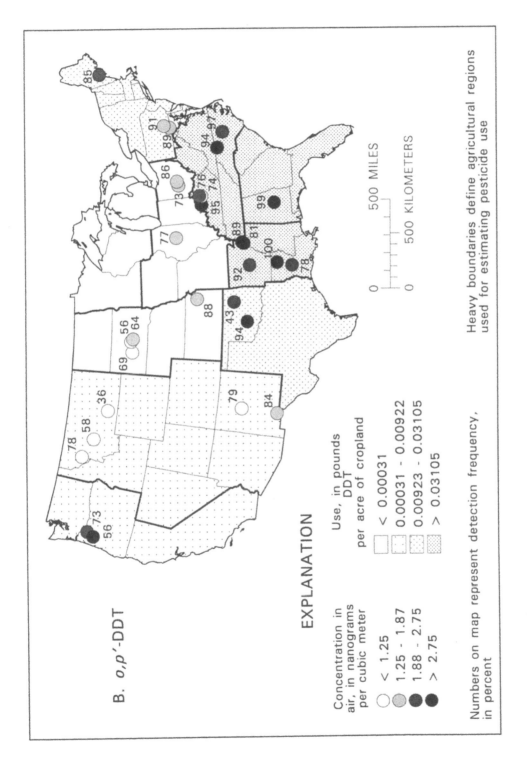

B. o,p'-DDT

EXPLANATION

Concentration in
air, in nanograms
per cubic meter

○ < 1.25
◔ 1.25 - 1.87
● 1.88 - 2.75
● > 2.75

Use, in pounds
DDT
per acre of cropland

☐ < 0.00031
☐ 0.00031 - 0.00922
☐ 0.00923 - 0.03105
☐ > 0.03105

Numbers on map represent detection frequency,
in percent

Heavy boundaries define agricultural regions
used for estimating pesticide use

0   500 MILES

0   500 KILOMETERS

FIGURE 3.4.--Continued

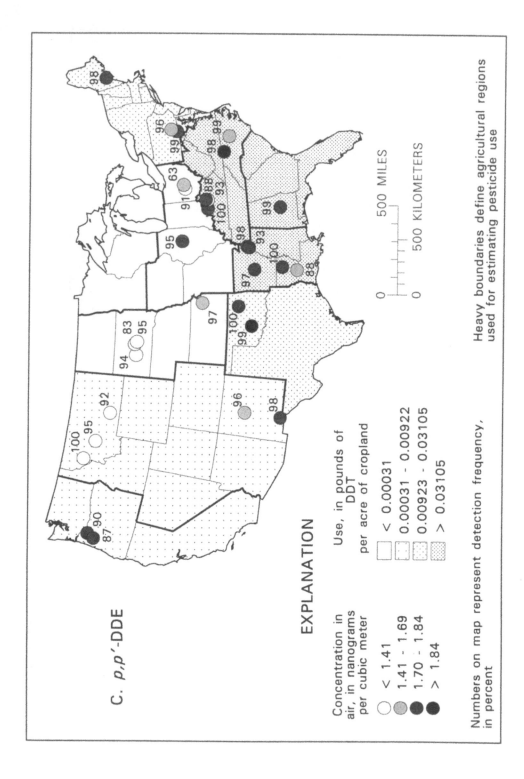

C. p,p'-DDE

EXPLANATION

Concentration in
air, in nanograms
per cubic meter

◯ < 1.41
◔ 1.41 - 1.69
● 1.70 - 1.84
● > 1.84

Use, in pounds of
DDT
per acre of cropland

☐ < 0.00031
▫ 0.00031 - 0.00922
▨ 0.00923 - 0.03105
▩ > 0.03105

Numbers on map represent detection frequency,
in percent

Heavy boundaries define agricultural regions
used for estimating pesticide use

0 ┤───┼───┼───┼───┤ 500 MILES

0 ┤───┼───┼───┼───┤ 500 KILOMETERS

FIGURE 3.4.—Continued

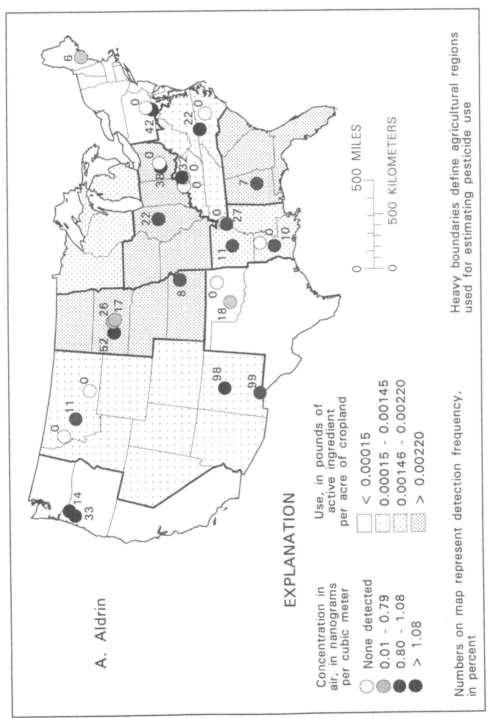

FIGURE 3.5. Average range of measured concentrations of (A) aldrin and (B) dieldrin in air and the detection frequency at each sampling site of Kutz and others (1976).

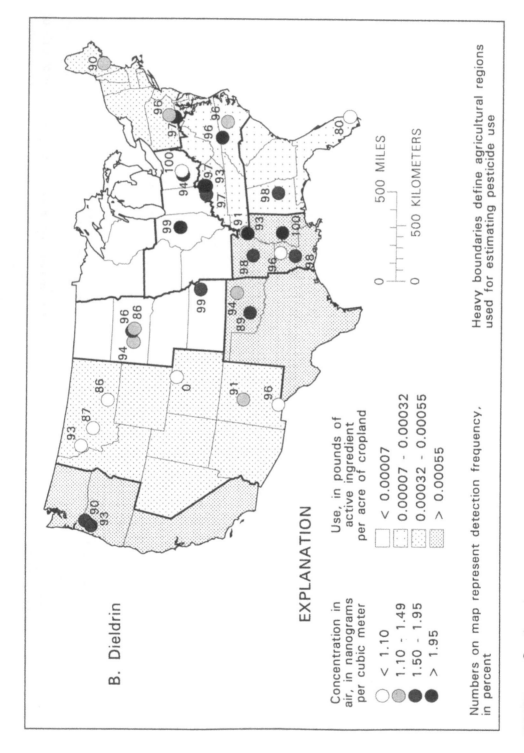

FIGURE 3.5.--Continued

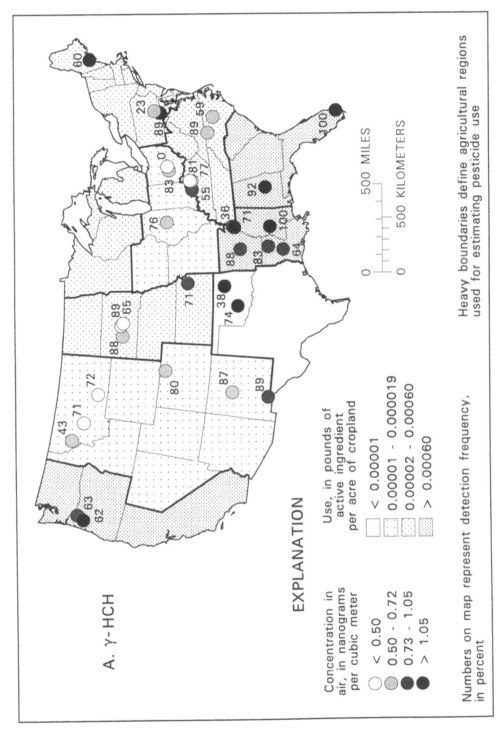

FIGURE 3.6. Average range of measured concentrations of (A) γ-HCH and (B) α-HCH in air and the detection frequency at each sampling site of Kutz and others (1976).

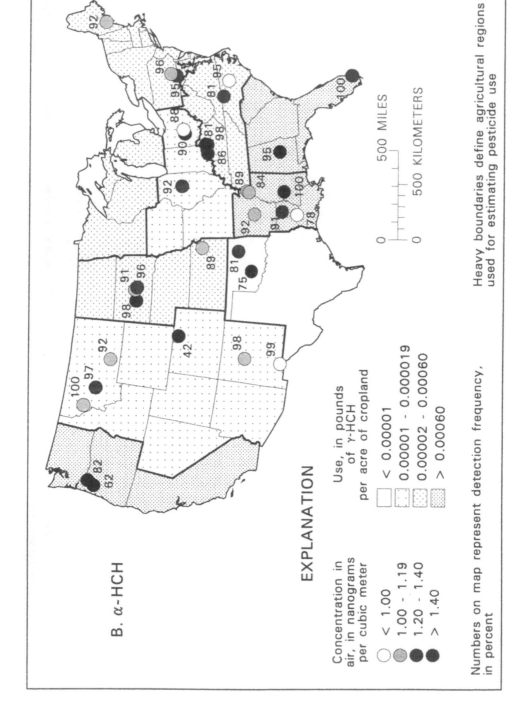

FIGURE 3.6.--Continued

as well as by the lumber industry, which may help explain the high concentrations measured in other areas of the country. α-HCH, the principal transformation product of γ-HCH, was also a component of technical lindane and has been found throughout the country at higher air concentrations and greater detection frequencies than γ-HCH.

The greatest intensity of studies that focused on organochlorine compounds in the atmosphere has taken place in the Great Lakes region since the 1960's (Strachan and Huneault, 1979; Eisenreich and others, 1981; Strachan, 1985; Eisenreich, 1987; Arimoto, 1989; Chan and Perkins, 1989; Voldner and Schroeder, 1989; Lane and others, 1992). These studies illustrated that atmospheric transport into the Great Lakes area, and deposition into the lakes was the primary input source of organic contaminants, including pesticides. Several studies (Rapaport and Eisenreich, 1986; Rice and others, 1986; Rapaport and Eisenreich, 1988; Voldner and Schroeder, 1989) have shown that toxaphene, an insecticide used heavily in the cotton-growing areas of the southern United States, can migrate northward and be deposited via the atmosphere into areas where use was limited or nonexistent.

ORGANOPHOSPHORUS INSECTICIDES

Organophosphorus compounds also have been heavily used for decades and many are still in high use. Although the actual amounts used in 1988 were slightly less than in 1971 (Table 3.1), they accounted for 65 percent of the insecticides used in agriculture. As a class, they are not as environmentally persistent as the organochlorine compounds. Because they are still widely used, they have been detected in the air and rain of many states (Figure 3.2B). Several organophosphorus insecticides were the focus of field worker reentry studies during the 1950's and 1960's, but often, they have not been included as target analytes.

Nationally, the organophosphorus compounds detected most often in air, rain, and fog were diazinon, methyl parathion, parathion, malathion, chlorpyrifos, and methidathion. Diazinon, methyl parathion, parathion, and malathion have been among the most heavily used insecticides in each of the last 3 decades, although diazinon, malathion, and parathion use has steadily declined during this time (Table 3.1). During the late 1960's, traces of several phosphorus and thiophosphorus compounds, and malathion were detected in the air of several agricultural communities in Florida, Georgia, Louisiana, Mississippi, and South Carolina (Tabor, 1965). Methyl parathion was detected in the air in three states (Alabama, Florida, and Mississippi) at concentrations ranging from 5.4 to 129 ng/m^3, and malathion was detected in Texas at 0.1 to 0.2 ng/m^3 (Tabor, 1965) and Florida at 2.0 ng/m^3 (Stanley and others, 1971). During the early 1970's, Kutz and others (1976) detected methyl parathion (0.3 to 42 ng/m^3), malathion (1.0 to 513 ng/m^3), and diazinon (0.6 to 7.3 ng/m^3) throughout the United States, but parathion was detected mainly in the southeastern United States at concentrations ranging from 1.1 to 239 ng/m^3. Since then, these compounds have been primarily analyzed for and detected in California fog and air (Glotfelty and others, 1987; Glotfelty and others, 1990a; Schomburg and others, 1991; Seiber and others, 1993).

Figure 3.7A shows the relationship between detection frequency in air and 1971 agricultural use for those organophosphorus insecticides analyzed for at 10 or more sites. This figure shows that there is no apparent national relationship. All of the organophosphorus insecticides in Figure 3.7A, with the exception of azodrin, leptophos, and phorate, were also detected in California although these three insecticides were not analyzed for in California. Of these 11 compounds, methidathion, chlorpyrifos, diazinon-OA, and parathion-OA (-OA is the oxygen analog transformation product of the parent compound) were detected only in California; however, the only site where these compounds were analyzed for outside of California was in Maryland (Glotfelty and others, 1987). When 1988 California use data for these compounds

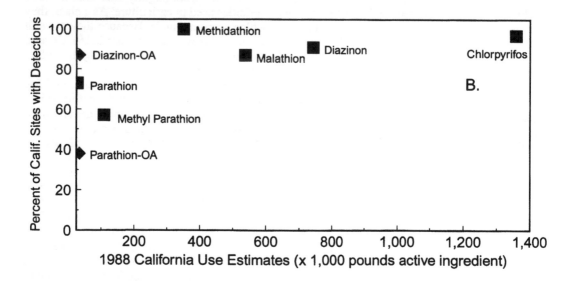

FIGURE 3.7. The relation between site detection frequency and agricultural use for organophosphorus insecticides and degradation products detected in air (A) nationally and in (B) California. Detection frequency data for *A* is for those compounds analyzed for at 10 or more sites. Use data for *A* is from the 1971 U.S. Department of Agriculture-National Agricultural Statistics Service agricultural pesticide use estimates (Andrilenas, 1974). Detection frequency data for *B* include all sites sampled in California except for field volatility studies. Use data for *B* are from Gianessi and Puffer (1992b). ■ denotes pesticide active ingredient and ◆ denotes pesticide degradation product.

(Gianessi and Puffer, 1992b) are compared to their California detection frequency (Figure 3.7B), a closer relationship is apparent. Azodrin and leptophos were not used in California in 1988 and phorate was not analyzed for in any study done in California so these three compounds are not included in Figure 3.7B.

All of the organophosphorus insecticides in Figure 3.7B were detected at greater than 50 percent of the sites sampled. The low methyl parathion and high methidathion detection frequencies relative to use, with respect to the other insecticides in the figure, can be explained by their use patterns and the types of studies that included them as target analytes. The primary use for methyl parathion in California is on rice (Gianessi and Puffer, 1992b), which is grown mainly in the north-central part of the state (California Department of Pesticide Regulations, 1990). The majority of the studies were done in the southern part of the state in orchards and other areas far removed from the rice-growing region. Seiber and others (1989) found high concentrations of methyl parathion as well as molinate and thiobencarb in the air sampled at four communities in an intensive rice-growing area of northern California. They found a correlation between the concentration in air and the applications of these three pesticides. One site not located near any rice production also was sampled and none of the three pesticides were detected at concentrations greater than the analytical limits of detection. Methidathion is primarily used on orchards, and the majority of studies that analyzed for it were done in or near orchards. Phorate is primarily used on row crops throughout the state, and in greater quantities than either parathion or methyl parathion. It has not been included as a target analyte in any study done in California.

Oudiz and Klein (1988) found an apparent correlation between county-wide parathion use and atmospheric concentrations in the San Joaquin Valley in central California and the Imperial Valley in southern California. Parathion use in Fresno County in the central San Joaquin Valley was about five times that of Kern County in the southern San Joaquin Valley during January 1986. The corresponding ratio of averaged county concentrations in air was about 6 to 1. Parathion use in Imperial County in the Imperial Valley during late September through October was slightly higher than that in Kern County, and the average concentrations in air were nearly equivalent to those in the southern San Joaquin Valley (Kern County) in January. These results indicate that detected pesticide concentrations in air can be directly correlated to use in the area.

Malathion, diazinon, and chlorpyrifos are all used on a variety of crops and orchards, while parathion is primarily used on row crops. A broad mixture of study types included these compounds as target analytes that had sampling sites in residential areas, in field crop areas, and in orchards. Together, the variety of studies, sampling locations, and different organophosphorus pesticides analyzed for throughout California combine to show a good correlation between pesticide use and occurrence in air.

Very few studies have analyzed for organophosphorus compounds in states other than California. Those studies that have analyzed for these compounds often focused on air and fog, and very few have investigated the organophosphorus pesticide content of rain and snow. Richards and others (1987) detected fonofos (<0.1 to 0.5 µg/L) in Indiana and Ohio rain. Malathion was recently detected in Iowa precipitation at 0.17 µg/L along with dimethoate (0.19 µg/L), fonofos (0.12 µg/L), and methyl parathion (1.60 µg/L) (Nations and Hallberg, 1992).

The organophosphorus insecticides from sites in the study by Kutz and others (1976) that had detection frequencies of 10 percent or more were diazinon (53 percent), malathion (19 percent), and methyl parathion (10 percent). Figure 3.8 shows the geographical distribution of the average air concentration and detection frequency at each site for these three insecticides. Also shown is the 1971 regional use for each insecticide (Andrilenas, 1974).

There is an apparent relation between clusters of sites with the highest air concentrations, regional use, and cropping patterns for each of these three insecticides. For example, the highest diazinon air concentrations are clustered in the Corn Belt and Appalachian regions (Figure 3.8A). Diazinon use in 1971 in the Corn Belt was more than 1 million lb a.i., and in the Appalachian region was more than 300,000 lb a.i. The primary use for diazinon was on corn and tobacco. Tobacco is a dominant crop in the Appalachian region and corn is also grown there, plus many of the sampling sites bordered the Corn Belt. There was very little reported diazinon use on wheat, cotton, and soybeans, all of which are grown in various regions throughout the country. The high air concentrations observed at the other sites where its reported agricultural use was low cannot be adequately explained given the large areal scale of this use data. It should be noted that diazinon has a high home and garden use, which may explain, in part, the high observed air concentrations in low agricultural use areas.

Total malathion use was only slightly less than diazinon in 1971, and the measured air concentrations were similar (Figure 3.8B). Malathion was primarily used on cotton and tobacco, which are extensively grown in the Delta, Southeastern, and Appalachian regions. The amount applied in the Delta region was low (Figure 3.8B), but the cropping patterns in this area may help explain the high observed air concentrations if the sampling sites were near cotton or tobacco fields. Malathion was also used on soybeans, corn, and, to a lesser extent, on wheat. The high average air concentrations observed in the Mountain and Northern Plains regions may be explained by the cropping patterns and pesticide use near the sampling sites.

The primary 1971 methyl parathion use was on cotton and, to a lesser extent, on soybeans, wheat, and tobacco. These crops correspond to the high observed average air concentrations in the Delta, Southeast, and Appalachian regions. Methyl parathion was also used on corn, which could help explain the range of observed air concentrations in the Mountain and Northern Plains regions (Figure 3.8C).

One additional fact that may help explain the closer relation between the average air concentration at a site and regional use for these three insecticides is the detection frequency. Diazinon was detected in an average of 53 percent of the samples while malathion and methyl parathion were only detected in an average of 19 and 10 percent of the samples, respectively. A high detection frequency indicates that the use of the compound is widespread or fairly constant throughout the sampling area. These data would be less influenced by one or two applications near the sampling site that would have resulted in a high averaged concentration, but low detection frequency.

OTHER INSECTICIDES

Only two carbamate insecticides reportedly have been analyzed for and detected in the atmosphere. Carbaryl was detected in California fog (0.069 to 4.0 µg/L) (Schomburg and others, 1991), and carbofuran was detected in Indiana, Ohio, West Virginia, and New York rain (less than 0.1 to 0.5 µg/L) (Richards and others, 1987). In general, insecticides other than the organochlorine and organophosphorus compounds have received little attention even though they comprise greater than 30 percent of all the insecticides currently used (Table 3.1).

TRIAZINE AND ACETANILIDE HERBICIDES

Triazine herbicides, which include atrazine, simazine, and cyanazine, and the acetanilide herbicides, which include alachlor and metolachlor, are used extensively in corn and sorghum production. As a class, triazine herbicide use has remained fairly steady at about 23 percent

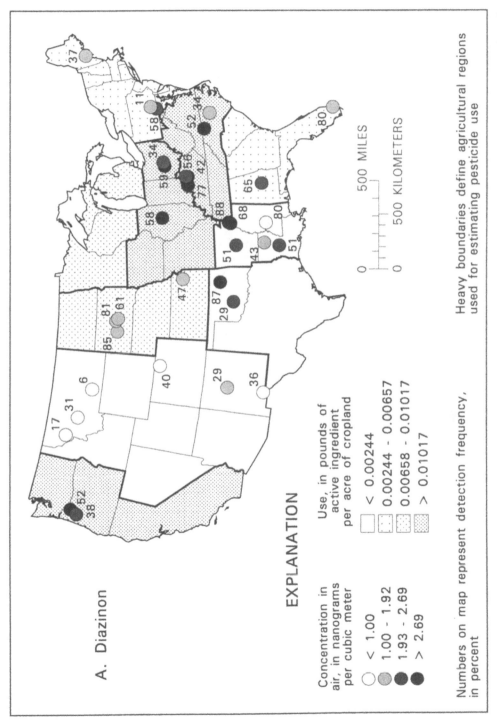

FIGURE 3.8. Average range of measured concentrations of (A) diazinon, (B) malathion, and (C) methyl parathion in air and the detection frequency at each sampling site of Kutz and others (1976).

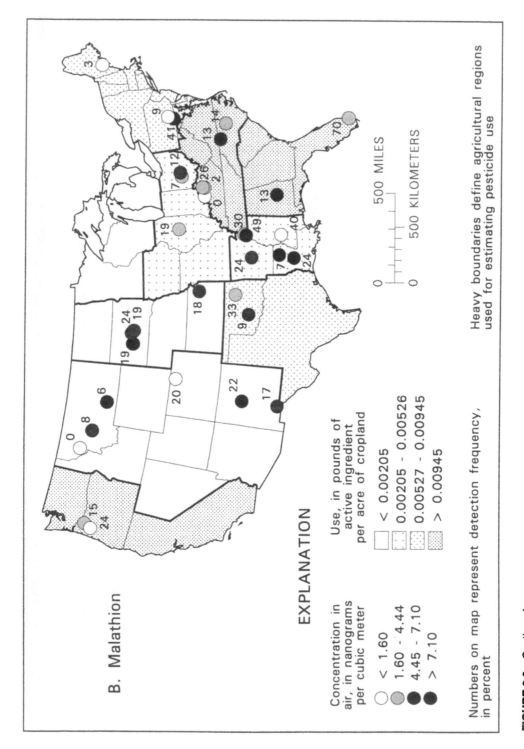

B. Malathion

EXPLANATION

Concentration in
air, in nanograms
per cubic meter

○ < 1.60
◐ 1.60 - 4.44
● 4.45 - 7.10
● > 7.10

Use, in pounds of
active ingredient
per acre of cropland

☐ < 0.00205
▦ 0.00205 - 0.00526
▦ 0.00527 - 0.00945
▦ > 0.00945

Numbers on map represent detection frequency,
in percent

Heavy boundaries define agricultural regions
used for estimating pesticide use

FIGURE 3.8.--Continued

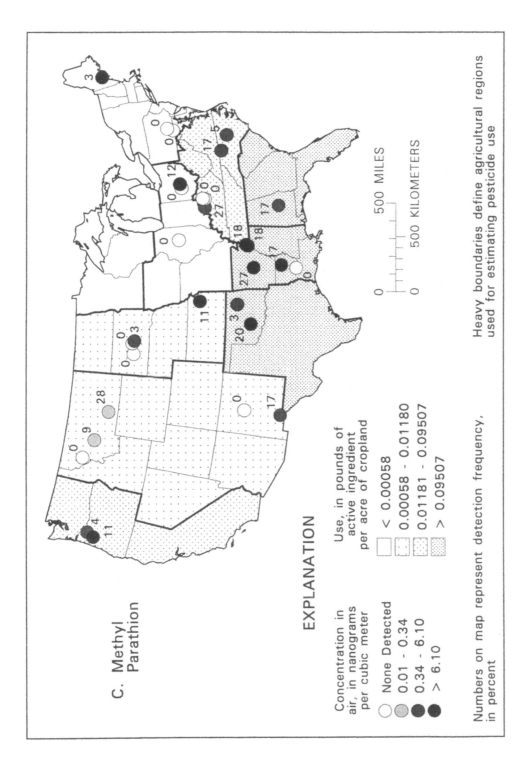

C. Methyl
Parathion

EXPLANATION

Concentration in
air, in nanograms
per cubic meter

○ None Detected
◍ 0.01 - 0.34
◍ 0.34 - 6.10
● > 6.10

Use, in pounds of
active ingredient
per acre of cropland

☐ < 0.00058
▫ 0.00058 - 0.01180
▨ 0.01181 - 0.09507
▨ > 0.09507

Numbers on map represent detection frequency,
in percent

Heavy boundaries define agricultural regions
used for estimating pesticide use

0   500 MILES

0   500 KILOMETERS

FIGURE 3.8.--*Continued*

(Table 3.1) with respect to total agricultural herbicide use. The amounts used, however, have steadily increased from 24 million lb a.i. in 1966 to nearly 104 million lb a.i. in 1988. Figure 3.9 shows that there is a good relation between national triazine and acetanilide herbicide use and detection frequency in rain. The majority of these data came from a 1990-91 study by Goolsby and others (1994) of weekly samples collected at 81 sites in the Midwestern and Northeastern United States. The one outlier, simazine, may result from significant nonagricultural use. The USEPA noncropland use estimate for simazine was 1.9 to 3.3 million lb a.i. for 1988 (Gianessi and Puffer, 1990). This amount is 49 to 83 percent of its agricultural use.

Several studies have analyzed for atrazine in air, but most included samples taken in conjunction with fog water samples (Glotfelty and others, 1987; Glotfelty and others, 1990a; Schomburg and others, 1991) or rain samples (Glotfelty and others, 1990c) and involved less than 10 sampling sites.

Although triazine herbicides have been in use since the 1960's, studies looking for these compounds in an atmospheric compartment did not begin until the late 1970's when Wu (1981) detected atrazine in Maryland precipitation (3 to 2,190 ng/L). Subsequent studies have focused on precipitation, and one or more triazine herbicides were detected in Maryland (0.031 to 0.48 ng/L) (Glotfelty and others, 1990c); Indiana (100 to greater than 1,000 ng/L); Ohio (less than 100 to greater than 1,000 ng/L); West Virginia (less than 100 to 500 ng/L); New York (less than 100 to greater than 1,000 ng/L) (Richards and others, 1987); Iowa (910 ng/L) (Nations and Hallberg, 1992); and Minnesota (less than 20 to 1,500 ng/L) (Capel, 1991).

Goolsby and others (1994) frequently detected atrazine and one of its major metabolites, deethylatrazine--and to a lesser extent cyanazine and simazine--in rain throughout the midwestern and northeastern United States. Figure 3.10A shows the 1988 atrazine use throughout the midwestern and northeastern United States. Figures 3.10B and C show the precipitation-weighted concentrations of atrazine for the same area for April through July, 1990

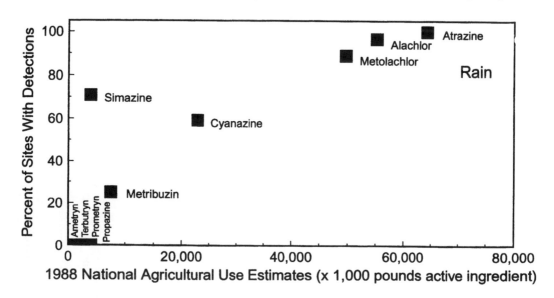

FIGURE 3.9. The relation between site detection frequency and national agricultural use for triazine and acetanilide herbicides detected in rain nationally. Detection frequency data are for those compounds analyzed at 10 or more sites, and the use data are from Gianessi and Puffer (1990).

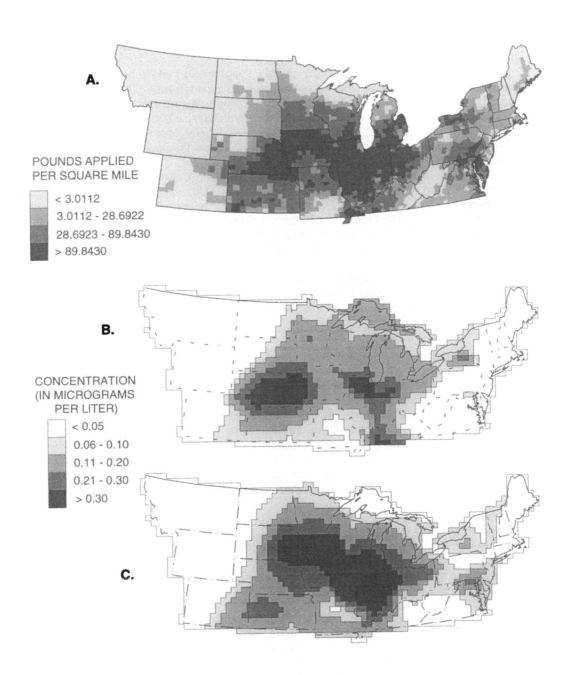

FIGURE 3.10. Atrazine use (A) in 1988 throughout the study area and precipitation-weighted concentrations throughout the midwestern and northeastern United States mid-April through mid-July, for (B) 1990 and (C) 1991.

and 1991 (from Goolsby and others, 1994). In terms of national use, over 77 percent of total atrazine use occurred in Illinois, Nebraska, Indiana, Iowa, Kansas, and Ohio with greater than 95 percent being used on corn and sorghum. These two figures show a very detailed and strong relation between atrazine use and detections in rain.

The acetanilide herbicides, which include alachlor and metolachlor, comprised about 26 percent of total herbicide use in 1988, up from only 5 percent in 1966. The actual amount used increased 20 times from less than 6 million lb a.i. in 1966 to nearly 119 million lb a.i. in 1988. These compounds are commonly used in conjunction with triazine herbicides and are commonly included in the same sample analysis. Only two acetanilide herbicides were analyzed for at 10 or more sites and are included in Figure 3.9. They appear to follow the same positive relation of use-to-detection frequency as the triazine herbicides.

Alachlor and metolachlor have been detected in the air (0.06 to 7.3 ng/m^3 and 0.07 to 1.7 ng/m^3, respectively) in Maryland (Glotfelty and others, 1990c) and in precipitation throughout the north-central and northeastern United States (Richards and others, 1987; Capel, 1991; Nations and Hallberg, 1992; Goolsby and others, 1994) in concentrations ranging from less than 100 to 21,000 ng/L for alachlor and from less than 100 to greater than 1,000 ng/L for metolachlor. Alachlor and atrazine dominated the other herbicides in detection frequency and concentration maximums.

Table 3.1 shows that triazine and acetanilide herbicides account for about 47 percent of total herbicide use in the United States during 1988. The remaining 53 percent is comprised of various other herbicide classes including chlorophenoxy acid, dinitrotoluidine, and thiocarbamate compounds.

OTHER HERBICIDES

Many types of herbicides other than the triazines and acetanilides are used in agriculture. Many of them have been detected in air and rain throughout the United States and elsewhere. Most of these herbicides fall into several classes including the chlorophenoxy acids, the thiocarbamates, and the dinitrotoluidines. Table 3.3 lists these other herbicides along with the state and year they were detected. Figure 3.11 shows the relation between detection frequency of other herbicides at 10 or more sites in air and their national use in agriculture. There appears to be a good correlation between detection frequency and use, but very few compounds are included in the figure and most of the data for this figure come from only one study (Kutz and others, 1976). Figure 3.11 illustrates two important points. The first is that while the number of other herbicides in current use in the United States is substantially greater than the number of triazine and acetanilides herbicides, the number of studies investigating their occurrence and distribution is very limited. The second is that much of the data for other herbicide occurrence in air comes from one study done in the early 1970's, and there is very little data on the national occurrence and distribution of other herbicides in rain.

The chlorophenoxy herbicides, which include 2,4-D and related esters, 2,4-DB, MCPA, and 2,4,5-T have been in use since the 1960's (Table 3.1). 2,4-D and its related esters have been extensively studied in air (10 to 13,500 ng/m^3) in Saskatchewan, Canada (Que Hee and others, 1975; Grover and others, 1976) and Washington (160 to 1,410 ng/m^3) (Farwell and others, 1976; Reisinger and Robinson, 1976) during the late 1960's and early 1970's. They also were detected, but not frequently, in precipitation at two sites in California (less than 10 to 80 ng/m^3) during the early 1980's (Shulters and others, 1987). Three 2,4-D esters were detected in the air of many of the 19 states (5.6 to 59 ng/m^3) in which Kutz and others (1976) sampled. In addition, two esters of 2,4,5-T were detected in Illinois and Tennessee, each at about 25 ng/m^3. Most studies that

TABLE 3.3. Herbicides other than triazines and acetanilide types that have been detected in air or rain, and the state and year in which they were detected

Compound	State	Year	Reference
Butylate	Iowa	1987-90	Nations and Hallberg, 1992
	Indiana, Ohio	1985	Richards and others, 1987
Dacthal	Illinois, Kansas, New Mexico, Oklahoma	1970-72	Kutz and others, 1976
DEF/Folex	California	1985	Glotfelty and others, 1987
	Mississippi	1967-68; 1972-74	Stanley and others, 1971; Arthur and others, 1976
	Tennessee	1970-72	Kutz and others, 1976
EPTC	Iowa	1987-90	Nations and Hallberg, 1992
	Indiana	1985	Richards and others, 1987
Molinate	California	1986	Seiber and others, 1989
Pendimethalin	California	1985	Glotfelty and others, 1987
	Iowa	1987-90	Nations and Hallberg, 1992
	Ohio	1985	Richards and others, 1987
Thiobencarb	California	1986	Seiber and others, 1989
Trifluralin	Arkansas, Illinois, Kansas, Kentucky, Louisiana, Maine, North Carolina, Ohio, Oklahoma, Tennessee	1970-72	Kutz and others, 1976
	Iowa	1987-90	Nations and Hallberg, 1992
2,4-D and related esters	Washington	1973, 1974	Farwell and others, 1976; Reisinger and Robinson, 1976
	California	1981-83	Shulters and others, 1987
	Utah	1967-68	Stanley and others, 1971
	Alabama, Arkansas, Illinois, Kansas, Kentucky, Louisiana, Maine, Montana, New Mexico, North Carolina, Ohio, Oklahoma, Oregon, Pennsylvania, South Dakota, Tennessee	1970-72	Kutz and others, 1976
2,4,5-T and related esters	Illinois, Oklahoma, Oregon, Tennessee	1970-72	Kutz and others, 1976

included chlorophenoxy herbicides took place during the late 1960's and early 1970's, and very few have been done since.

Kutz and others (1976) detected only the butoxyethanol ester of 2,4-D (14 percent) at frequencies of greater than 10 percent. Figure 3.12 shows its geographical distribution and detection frequency along with the regional agricultural use data in terms of total 2,4-D use in 1971. Individual 2,4-D ester use data are unavailable. Because the butoxyethanol ester was detected most frequently and at the highest concentrations relative to the other esters that were analyzed for by Kutz and others (1976), it was assumed that this ester was used in the greatest quantities. Total 2,4-D use in 1971 was second only to atrazine. The greatest amounts applied in terms of mass of 2,4-D per unit area of cropland occurred in the Northern Plains, Mountain, and Pacific regions. It was also used in large quantities in the Corn Belt and the Lake States. It was primarily used on corn, wheat, and other grains, as well as on pasture and rangeland. The concentration and detection frequency patterns of 2,4-D butoxyethanol ester in Figure 3.12 do not show any discernible relation to total 2,4-D use. This could be due to the fact that total 2,4-D

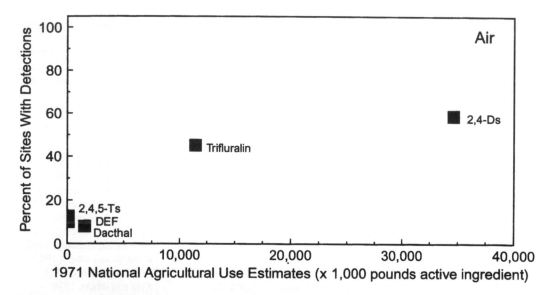

FIGURE 3.11. The relation between site detection frequency and national agricultural use for herbicides other than the triazine and acetanilide herbicides detected in air. Detection frequency data is for those compounds analyzed for at 10 or more sites, and use data is from the 1971 U.S. Department of Agriculture-National Agricultural Statistics Service agricultural pesticide use estimates (Andrilenas, 1974).

use does not approximate well the 2,4-D butoxyethanol ester use. Also, it is not clear whether the three 2,4-D esters detected by Kutz and others (1976) were the only ones analyzed for, or if the sampling method was able to collect the nonester form of 2,4-D. 2,4-D is formulated in many different ester forms.

Thiocarbamate herbicides include butylate, EPTC, molinate, and thiobencarb. Butylate and EPTC have been used extensively since the 1970's and were among the top 10 pesticides used in 1988 (Table 3.1). Both herbicides were detected in Iowa and Indiana rain at concentrations of less than 100 to 2,800 ng/L (Richards and others, 1987; Nations and Hallberg, 1992). Butylate also was detected in Ohio rain (less than 100 to 500 ng/L) (Richards and others, 1987). Molinate and thiobencarb primarily are used in rice production and were detected in air near rice-growing areas in California (2 to 630 ng/m^3) (Seiber and others, 1989).

The dinitrotoluidine herbicides include pendimethalin and trifluralin, which also are among the top 10 pesticides used in 1988 (Table 3.1). Trifluralin has been extensively used in wheat production since the 1960's and has been detected in air (0.7 to 10 ng/m^3) in a number of states (Kutz and others, 1976) and in Saskatchewan, Canada (Grover and others, 1988a). It also was detected in Iowa rain (970 ng/L) (Nations and Hallberg, 1992). Pendimethalin also was detected in Iowa and Ohio rain (less than 100 to 500 ng/L) (Richards and others, 1987; Nations and Hallberg, 1992) as well as in California air and fog (Glotfelty and others, 1987).

Dacthal is a general use, broadleaf herbicide that was detected in air in Illinois, Kansas, Oklahoma, and New Mexico during the early 1970's (0.5 to 1.1 ng/m^3) (Kutz and others, 1976). DEF and folex are similar organophosphorus herbicides used primarily as cotton defoliants. They have been detected in California air (0.03 to 0.14 ng/m^3) and fog (250 to 800 ng/L). (Glotfelty and others, 1987) and in Mississippi air (0.1 to 16 ng/m^3) (Stanley and others, 1971; Arthur and others, 1976; Kutz and others, 1976).

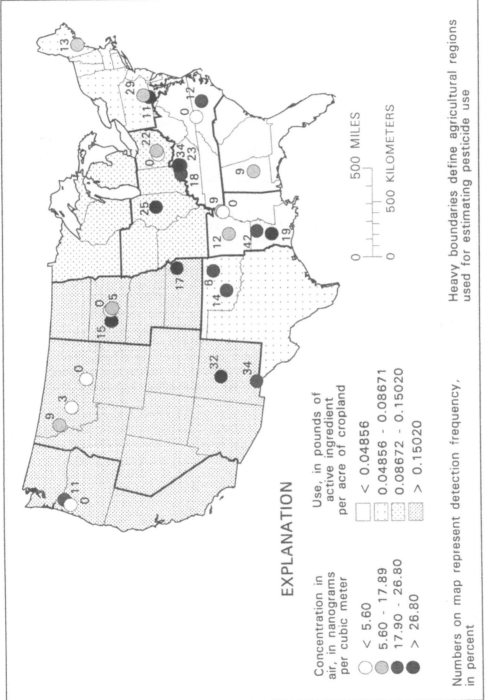

FIGURE 3.12. Average range of measured concentrations of 2,4-D in air and the detection frequency at each sampling site of Kutz and others (1976).

LONG-TERM TRENDS

Data are adequate for assessing long-term trends only for organochlorine compounds, and most data are for the Great Lakes region, Canada, and the Arctic. Ombrotrophic peatlands have been used to elucidate the historical use and atmospheric deposition of such organochlorine compounds as PCBs, DDTs, HCHs, HCB, and toxaphene (Rapaport and others, 1985; Rapaport and Eisenreich, 1986, 1988). Figure 3.13 shows that the atmospheric deposition of DDT closely followed its production and use. Larsson and Okla (1989) found that the dry deposition of DDT in Sweden decreased significantly between 1973 and 1985 in conjunction with the restrictions on its use. In contrast, PCB concentrations stayed nearly the same and were believed to be due to contamination by local combustion sources. Addison and Zinck (1986) found that the PCB concentration in male western Arctic ringed seal (*Phoca hispida*) blubber decreased significantly from 3.7 µg/g wet weight in 1972 to 1.3 µg/g wet weight in 1981. This decline coincided with the ban on PCB manufacturing and use in the United States and Canada in the early 1970's. In contrast, the DDT concentrations did not show any clear decline over the same time period. The ratio of *p,p'*-DDT to *p,p'*-DDE suggested that there was a continued fresh supply of DDT into the western Arctic. They speculated that the most probable route and source of this continuing supply of DDT was atmospheric transport from the Far East and Eurasia, where DDT and other organochlorine pesticides are still heavily used. The recent study of organochlorine pesticides, including DDT, over the world's oceans further supports this hypothesis (Iwata and others, 1993).

Agricultural use of γ-HCH dropped 10-fold from 1971 to 1988 (Table 3.1) in the United States and Canada and its measured atmospheric concentration has been declining for over a decade.   Figure 3.14 shows this decrease in average yearly precipitation concentrations at various sites across Canada. The higher concentrations of α-HCH result from the environmental transformation of γ-HCH. The α- isomer is also present as an impurity in technical lindane, which is still used in Central and South America, and Asia, and can be transported north via the atmosphere, although interhemispheric exchange is not believed to be significant (Ballschmiter and Wittlinger, 1991).

Other organochlorine pesticides such as aldrin, heptachlor, heptachlor epoxide, dieldrin, endrin, and mirex are not frequently detected in atmospheric samples taken in the United States and Canada. When they are detected, they are present in very low concentrations (Strachan and Huneault, 1979; Strachan, 1985; Chan and Perkins, 1989). They are also detected with varying frequency at low concentrations in the Arctic (Hargrave and others, 1988; Gregor and Gummer, 1989; Bidleman and others, 1990; Gregor, 1990) and no clear trends for these compounds are evident.

SUMMARY

An accurate assessment of the variety and extent of pesticides present in the atmosphere is difficult to make on the basis of information from existing studies. Only two national or large-scale regional studies have been done in the last 30 years (Kutz and others, 1976; Goolsby and others, 1994), and the nature of the data from each of these studies makes meaningful comparisons difficult. Kutz and others (1976) analyzed air samples for a wide variety of insecticides and herbicides throughout the country, but concentrated most of their sites in the eastern United States. They also focused on sites with high detection probabilities. In contrast, Goolsby and others (1994) analyzed rain samples only for triazine and acetanilide herbicides and concentrated their sampling effort on the Midwest and Northeast. They used existing National Atmospheric Deposition Program/National Trends Network (NADP/NTN) sampling sites that

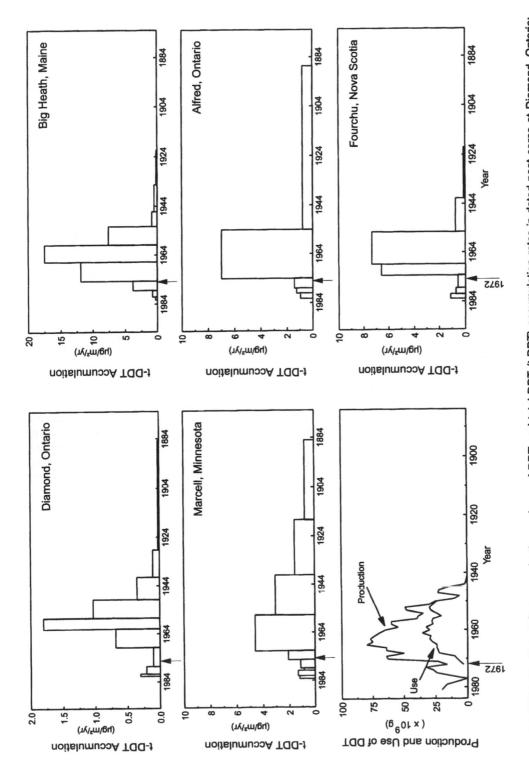

FIGURE 3.13. United States production and use of DDT and total DDT (t-DDT) accumulation rates in dated peat cores at Diamond, Ontario; Marcell, Minnesota; Big Heath, Maine; Alfred, Ontario; and Fourchu, Nova Scotia (adapted from Rapaport and others, 1985).

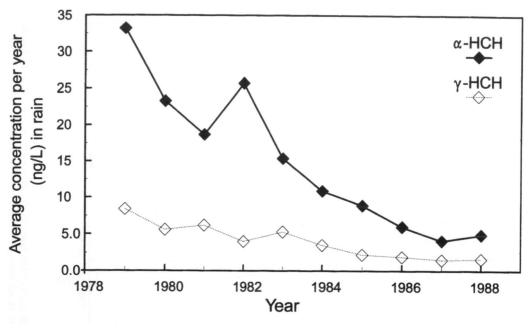

FIGURE 3.14. Average HCH concentrations (α and γ) in precipitation at various sites across Canada. Data from Brooksbank, 1983; Strachan, 1988; Chan and Perkins, 1989; and Brun and others, 1991.

were selected to represent major physiographic, agricultural, aquatic, and forested areas within each state, region, or ecoregion (Bigelow, 1984).

Combined results from the state and local and the national and multistate monitoring studies listed in Tables 2.3 and 2.4 indicate that a wide variety of pesticides are present in the air, rain, snow, and fog. How and why they are there will be explained in the next sections, but the overall conclusions are that nearly every pesticide that has been analyzed for has been detected in one or more atmospheric matrices throughout the country at different times of the year. There is ample evidence that pesticides used in one area of the country are transported through the atmosphere and are deposited in other areas of the country, sometimes in areas where pesticides are not used.

Relations between pesticide use and detected concentrations and frequencies are not clearly defined from the combined data in Tables 2.3 and 2.4. Many of the discrepancies, such as high detection frequency but low use, or high use but low detection frequency, however, can be explained by the physical placement of the sampling sites, the analytical detection limitations, or the persistence or degradation of the parent compound. The relation between pesticide use and detected concentrations and frequencies improved when selective data from the 1980's were used, as shown by the California studies on organophosphorus insecticides in air and by the study by Goolsby and others (1994) on the triazine and acetanilide herbicides in rain.

To determine the extent and magnitude of water-quality and human health effects, if any, from atmospheric deposition of pesticides, the long-term distribution and trends of these chemicals in the atmosphere need to be assessed. To this end, long-term regional or national-scale studies are needed to analyze for a representative set of current, high-use pesticides with uniformity of sampling protocol and analytical techniques.

CHAPTER 4

Governing Processes

An understanding of the occurrence and distribution of pesticides in the atmosphere requires consideration of pesticide sources, transport processes, and mechanisms of transformation and removal from the atmosphere. The following chapter is an overview of these factors and provides a background for the subsequent, more detailed analysis of specific key topics about pesticides in the atmosphere.

4.1 SOURCES

The greatest source of pesticide contamination of the atmosphere is agricultural use, which involves vast acreage and the use of millions of pounds of chemicals yearly. About 75 percent of the pesticides used annually are on agricultural crops (Aspelin and others, 1992; Aspelin, 1994). Other sources of pesticide contamination of the atmosphere include manufacturing processes and waste effluents, urban, industrial, and right-of-way weed control, turf management of golf courses, parks, and cemeteries, and large-scale aerial spraying for the abatement of pests such as mosquitoes, the Mediterranean fruit fly, the gypsy moth, and the Japanese beetle. Although total agricultural use of pesticides is greater than urban use because of the larger area, the intensity of urban use (mass per unit area) has been estimated to be equivalent to that used by farmers (Farm Chemicals, 1992; Gold and Groffman, 1993).

Because pesticides are primarily used in agriculture which involves large acreage, large quantities, and most major types of pesticides, the focus of this section is on agricultural sources and related processes. The processes described, however, are also applicable to the other sources mentioned above. The most important agricultural sources fall into two main categories: application and post-application processes.

APPLICATION PROCESSES

Off-target drift during pesticide application occurs to varying degrees, ranging from 1 to 75 percent of the applied spray (Grover and others, 1972; Yates and Akesson, 1973; Nordby and Skuterud, 1975; Farwell and others, 1976; White and others, 1977; Grover and others, 1978, 1985, 1988b; Cliath and others, 1980; Willis and others, 1983; Clendening and others, 1990). A portion of the off-target drift usually is deposited quickly within a short distance of the application site, but some remains airborne longer, returns slowly to the surface, and can be carried longer distances downwind. Many different factors combine to affect drift behavior during the application process and the rate of off-target deposition. Three main categories of factors are application methods, formulations, and spray-cloud processes.

Application Methods

A uniform distribution is the goal for most pesticide applications. Herbicides commonly are directed at any part of the unwanted plant, whereas insecticides and fungicides ideally are directed at microhabitats within the foliage canopy (Himel and others, 1990). Various pesticide application systems include ground-rig broadcast sprayers, aerial methods, and orchard misters. The potential for drift and volatilization during application generally increases with each of these methods, respectively. Ground-rig broadcast sprays are generally directed toward the ground as are aerial application methods. Aerial methods, however, have higher drift and volatilization potentials than ground rigs given the same droplet size distribution. Air currents produced by the aircraft have a major effect on the trajectories of the fine particles released and can increase their drift potential. In general, spray drift from aerial applications is about five times greater than from ground-rig applications (Ware and others, 1969; Medved, 1975). Orchard radial and axial fan mist-blowers direct the spray up and away from the ground in an effort to cover the entire tree or crop canopy. Drift from this type of application has been measured at distances of up to six times greater than from aerial applications (Ware and others, 1969; Frost and Ware, 1970). Pesticides also can be added to irrigation water. This technique, called chemigation, can be used in flood, drip, and overhead sprinkler irrigation systems.

Formulations

Many different types of pesticide carrier formulations exist, and diluents range from water, various solvents, surfactants, and oils, to chalk, clays, ground walnut shells, and so forth. The use of any particular formulation and carrier is dependent on the required action and placement of the pesticide. Emulsifiable concentrates are currently extensively used because they are easy to apply with modern spray equipment and water as the typical diluent. Other formulations include flowable and wettable powders, which are finely ground dry formulations and active ingredients suspended in a liquid, usually water. Granular formulations and pellets come in various sizes ($\approx$250 to 2,500 μm diameter) and disintegration or release properties. They usually do not need a water carrier or dispersant and are often ready-made for application. Dust formulations (5 to 20 μm diameter) can penetrate dense canopies, but are easily carried off-target by wind. Plastic or starch micro-encapsulated formulations are used for time release of the chemical. Gases (methyl bromide, ethylene oxide) and very volatile liquids (ethylene dibromide, carbon disulfide, dichloropropene) are commonly used in preplant fumigation of soil and usually are injected into the soil. These compounds are extremely volatile and one of their primary dissipation routes is by volatilization into the atmosphere if they are not contained (Roberts and Stoydin, 1976; Majewski and others, 1995), although little environmental fate information is currently available in the literature.

Actual application rates depend on the pesticide being used. They range from ultra-low volume at less than 2 L/ha, to high volume at greater than 300 L/ha. If the spray droplets are small or if appreciable volatilization of the carrier liquid occurs, the droplets, dust, or powder particulates can become suspended in air. These small droplets and particles have low depositional velocities and are more likely to be carried off-target by even a slight wind. Drift potential during application is usually very low with granular formulations. In contrast, dusts have a very high drift potential when used with conventional applicators (Yates and Akesson, 1973).

The only major influence on the size of a droplet after it has been formed by the spray nozzle is volatilization. Evaporation of spray droplets and the associated pesticide can occur as they travel from the nozzle to the ground. Evaporation of oil-water pesticide emulsion droplets is about the same as for pure water droplets (Yates and Akesson, 1973), and highly dilute aqueous spray droplets of less than 150 µm diameter evaporate rapidly (Spillman, 1984). Under atmospheric conditions common during pesticide application, greater than 40 percent of the original spray volume can be lost by evaporation before impact (Cunningham and others, 1962). The droplet size reduction due to evaporation can result in the finer droplets of a normal distribution disappearing while the larger drops are reduced in size. Formulating agents are sometimes added to decrease the vapor pressure of the carrier, which reduces the evaporation rate and slows the reduction in droplet size.   The result is that the droplet itself may not disappear before reaching the ground, but the distribution of the smaller diameter droplets, their concentration, their overall flight time, and the off-target drift potential can increase. Wetting agents such as surfactants and oils reduce surface tension which increases droplet breakup and drift potential.

Spray-Cloud Processes

The behavior of a spray cloud is very complex and is influenced by atmospheric movements that are equally complex and difficult to explain thoroughly. The droplet size spectrum of the spray cloud is influenced by many of the same factors that affect drift during application (Coutts and Yates, 1968). A drifting spray cloud can spread horizontally and vertically down- and cross-wind. The larger droplets will rapidly settle to the ground while the finer ones with low settling velocities can remain airborne for longer periods of time and be carried appreciable distances downwind from the application site. The main parameters affecting the dispersion of the drifting cloud are wind speed and direction, ambient temperature and humidity, incoming solar radiation, and other micrometeorological parameters related to atmospheric stability; that is, the degree of turbulent mixing (Nordby and Skuterud, 1975).

The concentration and deposition of a drifting spray cloud is dependent on atmospheric diffusion, which is a function of the intensity and spectrum of atmospheric turbulence. There are two main types of atmospheric turbulence generated within the surface boundary layer: mechanical and thermal. The surface boundary layer is the lowest part of the atmosphere in direct contact with the surface. This is the zone in which the wind velocity and turbulence increase logarithmically with height above the surface until they reach some chosen fraction of magnitude of the free-moving airstream; for example, 99 percent. Mechanical turbulence is generated near the surface by the frictional and form drag forces at the surface and is related to the increase in wind speed with height. Thermal turbulence is generated by buoyant air movements induced by vertical temperature gradients (Monteith, 1973). High frequency, small air motion fluctuations primarily are due to mechanical turbulence, while low frequency, larger air motion fluctuations are the result of thermal turbulence (Rosenberg and others, 1983).   Turbulence is enhanced by buoyant forces under unstable conditions and is suppressed under stable conditions.

The increase in turbulence with height depends on the stability structure of the atmosphere. Air parcels displaced from one level to another transfer momentum to the surrounding air, which can either enhance or diminish turbulence. Large-scale eddies that are much larger than a drifting spray cloud, move the cloud downwind with little dispersion. Small-scale eddies that are much smaller than the drifting cloud, cause a slight growth in the cloud and a corresponding decrease in concentration due to mixing. Those eddies that are the same size as the drifting cloud can rapidly disperse it due to turbulent mixing (Christensen and others, 1969).

Transport of spray droplets to a surface is dependent on atmospheric turbulence and gravitational forces. Droplet size has a considerable effect on drift and evaporation. Turbulent influences are inversely proportional to the diameter of the droplet whereas gravitational forces are directly proportional to it. Small droplets are, therefore, primarily transported on turbulent eddies, and their impact on a target depends on their size, velocity, and target geometry. Fine particle sizes are dispersed better, but their deposition velocities and trajectories are more influenced by external factors such as the gustiness of the wind. Small droplets (less than 0.1 μm diameter) also have deposition velocities that are negligible compared to the atmosphere's turbulent motions. This means that gravitational settling will have less of an influence on them than atmospheric turbulence, and they will take a less direct path to the surface. Gravitational settling has no real influence on droplets of less than 100 μm diameter under most field spray conditions (Himel and others, 1990) whereas large droplets are primarily affected by gravity.

Spraying with large droplets increases the deposition accuracy, but the target coverage may not be sufficient, thereby necessitating greater application rates. Typical droplet diameters for most spray application conditions range between 200 and 300 μm. The upper limit of droplet diameter for drift concerns is about 100 μm (Cunningham and others, 1962).

The stability of the atmosphere has a significant effect on application spray drift, post-application volatilization rates, drift in terms of the downwind distance a vapor or aerosol cloud travels, and the concentration of the deposits. Unstable situations occur when the temperature of the surface is greater than the overlying air, resulting in rising heat plumes and dispersive turbulence. A stable or inversion atmosphere has no thermally induced vertical fluctuations, and very little vertical dispersion occurs. Stable conditions can result in high pollutant concentrations near the surface that can be maintained for long downwind distances. Long-range drift for all application systems can be reduced by spraying during calm (low wind speed), neutral atmospheric conditions. These conditions can be conducive to short-range drift and deposition, and buffer zones have been recommended to minimize short-range crop damage by drift (Payne, 1992; Payne and Thompson, 1992). Cooler ambient temperatures during application will also reduce drift by minimizing droplet evaporation.

POST-APPLICATION PROCESSES

Once on the target surface, the pesticide residue can volatilize by evaporation or sublimation or be transported into the atmosphere attached to dust particles (Spencer and others, 1984; Chyou and Sleicher, 1986; Glotfelty and others, 1989; Clendening and others, 1990; Grover, 1991). Tillage practices affect both of these processes. Post-application volatilization from treated fields represents a secondary form of off-target pesticide drift that takes place over a much longer time period. This volatilization is a continuous process, and the resulting drift can be a significant source of pesticide input into the lower atmosphere. Volatilization from soil and surface waters is a major dissipation route for many pesticides, and as much as 80 to 90 percent can be lost within a few days of application for certain compounds (Soderquist and others, 1977; Cliath and others, 1980; Glotfelty and others, 1984; Majewski and others, 1993; Majewski and others, 1995).

The volatilization rate from soil, water, and vegetative surface sources depends mainly on the chemical's effective vapor pressure at the surface and its rate of movement away from the surface (Spencer and Cliath, 1974; Spencer and others, 1982). However, these two factors can be influenced in a number of ways, including:

(1) Application and formulation type, and whether it is surface applied or incorporated;

    (2) Degree of sorption to the application surface; that is, the organic matter and clay content of soil, suspended biota and organic matter in water, and type and density of the vegetative surface, as well as the amount of surface waxes and oils on the leaves;

    (3) Soil moisture distribution and temperature;

    (4) Nature of the air-surface interface through which the chemical must pass;

    (5) Soil tillage practices such as conventional, low, or no-till; and

    (6) Micrometeorological conditions above the soil surface.

Volatilization usually follows diurnal cycles, and is very dependent on the solar energy input and the atmospheric stability. In general, the volatilization rate is proportional to the solar energy input and the atmospheric turbulence, both of which are typically maximized around solar noon and diminished at night. The nature of the surface also plays an important role in the volatilization process. For example, soil dries out with no additional moisture inputs, and the drying of even the top few millimeters of the surface has been shown to effectively suppress pesticide volatilization (Spencer and others, 1969; Harper and others, 1976; Grover and others, 1988a; Glotfelty and others, 1989; Majewski and others, 1991). For dry soils, the volatilization dependence on solar energy is reduced and is almost exclusively dependent on additional moisture inputs. In this situation, volatilization maxima occur with dew formation, usually in the early mornings and evenings, and with rain and irrigation (Cliath and others, 1980; Hollingsworth, 1980; Glotfelty and others, 1984; Grover and others, 1985; Majewski and others, 1990).

Incorporation of the pesticide into the top few centimeters of the soil can reduce the initial high volatilization losses during and immediately after the application (Spencer, 1987; Grover and others, 1988b). Even injecting pesticide formulations below the surface of water considerably reduces the volatilization rate over surface applications (Maguire, 1991). The total long-term volatility losses for injected and incorporated cases may be similar to the total surface-applied losses because the volatilization rates of the incorporated pesticide will be more constant over time, whereas the surface-applied pesticides have a very rapid initial loss that leaves less of the material at the surface, which, in turn, reduces the volatilization rate (Nash and Hill, 1990).

Pesticide volatilization from soil is complicated and many factors influence pesticide movement to and from the surface. Temperature can affect volatilization through its effect on vapor pressure. For incorporated chemicals, an increase in soil temperature may enhance their movement to the surface by diffusion, and by mass flow as water is pulled to the surface by the suction gradient created by its volatilization from the surface (Hartley, 1969; Spencer and Cliath, 1973). Water competes with and can displace bound pesticides from active soil adsorptive sites (Spencer and others, 1969; Spencer and Cliath, 1970). Through the upward movement and volatilization of water, pesticide residues can accumulate at the surface and result in an increase in volatilization rate. High temperatures can also decrease the evaporative rate by drying the soil surface as mentioned above. A high soil organic matter content enhances pesticide binding and reduces the volatilization rate. In moist soil situations, the additional partitioning between the soil particles and the surrounding water also must be considered. Table 4.1 shows examples of the volatilization rates for various pesticides and the differences between surface applications and incorporation.

TABLE 4.1. Volatilization losses for various pesticides after surface application or incorporation

[Data extracted from Table 2.2]

Compound	Application type	Loss by volatilization		Reference
		(%)	In days	
Alachlor	Surface applied	19	21	Glotfelty and others, 1989
Atrazine	Surface applied	2.4	21	Glotfelty and others, 1989
Chlordane	Surface applied	50	2.5	Glotfelty and others, 1984
	Surface applied	2	2.1	Glotfelty and others, 1984
Chlorpropham	Surface applied	15	9	Turner and others, 1978
Chlorpyrifos	Surface applied	0.2	4	Majewski and others, 1990
2,4-D (isooctyl ester)	Surface applied	20.8	5	Grover and others, 1985
Dacthal	Surface applied	2	1.4	Glotfelty and others, 1984
	Surface applied	10	21	Ross and others, 1990
	Surface applied	40	21	Majewski and others, 1991
DDT	Surface applied	65	10.3	Willis and others, 1983
Diazinon	Surface applied	0.2	4	Majewski and others, 1990
Eptam	Irrigation water	73.6	2.2	Cliath and others, 1980
Heptachlor	Surface applied	14-40	2.1	Glotfelty and others, 1984
	Surface applied	50	0.25	Glotfelty and others, 1984
	Surface applied	90	6	Glotfelty and others, 1984
HCH, γ-	Surface applied	12	2.1	Glotfelty and others, 1984
	Surface applied	50	0.25	Glotfelty and others, 1984
	Surface applied	6.6	4	Majewski and others, 1990
MCPA[1]	Surface applied	0.7	4	Seiber and others, 1986
Methyl bromide	Incorporated[2]	22	5	Majewski and others, 1995
	Incorporated	89	5	Majewski and others, 1995
Molinate[1]	Surface applied	35	4	Seiber and others, 1986
Molinate[1]	Surface applied	78	7	Soderquist and others, 1977
Nitrapyrin	Surface applied	5.5	4	Majewski and others, 1990
Simazine	Surface applied	1.3	21	Glotfelty and others, 1989
Thiobencarb[1]	Surface applied	1.6	4	Seiber and others, 1986
Toxaphene	Surface applied	31	21	Glotfelty and others, 1989
	Surface applied	50	80	Seiber and others, 1979
Toxaphene[3]	Surface applied	80	50	Seiber and others, 1979
Toxaphene	Surface applied	21	4.7	Willis and others, 1983
(Second application)	Surface applied	60	10.8	Willis and others, 1983
Triallate	Incorporated	15	30	Grover and others, 1988b
	Surface applied	74	5	Majewski and others, 1993
Trifluralin	Surface applied	2-25	2.1	Glotfelty and others, 1984
	Surface applied	50	0.13-0.31	Glotfelty and others, 1984
	Surface applied	90	2.5-7	Glotfelty and others, 1984
	Surface applied	54	5	Majewski and others, 1993
	Incorporated	20	30	Grover and others, 1988b
	Incorporated	25.9	120	White and others, 1977
	Incorporated	22	120	Harper and others, 1976

[1]Ricefield water
[2]Tarped field
[3]Cotton foliage

Wind Erosion

Wind erosion of formulation dusts, small granules, and pesticides bound to surface soil is another mechanism by which applied pesticides reach the atmosphere, although it is generally considered to be less important than volatilization (Glotfelty and others, 1989). Factors that influence the erodibility of soil include horizontal wind speed, precipitation, temperature, soil weathering, and cultivation practice (Chepil and Woodruff, 1963). Very large particles (500 to 1,000 μm diameter) tend to roll along the ground and, generally, do not become airborne, but they can break apart into smaller particles or dislodge small particles from the surface as they roll. Particles in the size range of 100 to 500 μm diameter move by saltation, a skipping action that is the most important process in terms of the wind erosion problem and in moving the greatest amount of soil when there is a long downwind fetch (Nicholson, 1988b). Although large and saltating size particles can move horizontally great distances, depending on the wind speed, their vertical movement is rarely above one meter (Anspaugh and others, 1975) and they are usually deposited near their source.

The most important particle size range, with respect to atmospheric chemistry and physics is 0.002 to 10 μm (Finlayson-Pitts and Pitts, 1986). Intermediate sized, or accumulation range particles (0.08 to 1-2 μm diameter) arise from condensation of low volatility vapors and coagulation of smaller particles.   Accumulation range particles are not affected by rapid gravitational settling and are only slowly removed by wet and dry deposition, therefore they are susceptible to long atmospheric lifetimes and have high potential for long-range atmospheric transport (Bidleman, 1988).   The smallest particles, known as transient or Aitken nuclei (less than 0.08 μm diameter) arise from ambient temperature gas-to-particle conversion and combustion processes in which hot, supersaturated vapors are formed and subsequently undergo condensation (Finlayson-Pitts and Pitts, 1986). The lifetimes of Aitken particles are short because they rapidly coagulate (Bidleman, 1988). There have been few field studies that measured the pesticide content of windblown soil, dust, and particulate matter from agricultural fields.

Tillage Practices

Tillage practices used to cultivate agricultural land can affect pesticide transport into the lower atmosphere by either volatilization or wind erosion. Doubling the soil organic matter content can cut volatilization rates by a factor of about 2, and a 2 to 10°C cooler soil surface temperature can reduce volatilization by as much as a factor of 2 to 4 (Spencer and others, 1973; Spencer, 1987). The degree of remaining plant residue (mulch) can change the microclimate at the soil surface, which affects the energy balance, moisture distribution, and rate of vapor exchange. The mulch insulates the soil and can result in a surface temperature that is 2 to 10°C cooler than bare soil (Glotfelty, 1987). Mulch improves water retention capabilities of the soil, which increases its thermal conductivity and allows heat to flow into the subsoil. It can decrease soil erosion and runoff, stabilize the organic matter content, lower the pH, and improve the soil structure (Glotfelty, 1987). Mulch also can change the surface albedo and reflect incoming radiation instead of absorbing it, which also cools the soil.

There are three basic types of tillage practices:

(1) Conventional tillage, where the soil is thoroughly mixed within the plow depth (the Ap horizon--a dark, uniform surface cap of about 15 to 25 cm in depth);

> (2) Conservation tillage, which leaves at least 30 percent of plant
>   residue covering the soil surface after planting; and
> (3) No-till, which leaves 90 to 100 percent residue cover.

Conventional tillage uniformly distributes crop residues, organic matter, available nitrogen, phosphorus, calcium, potassium, magnesium, pH, soil microorganisms and, in some cases, agricultural chemicals throughout the plow depth (Thomas and Frye, 1984). Conventional tillage also increases organic matter breakdown. In conventional tillage, pesticide volatilization is influenced by the properties of the soil such as organic matter and moisture content, and surface roughness as described above.

The remaining dead, surface plant material in conservation tillage and no-till forms a natural mulch resulting in conditions resembling a permanent pasture (also, see Wauchope, 1987). For the purposes of this review, those processes associated with no-till can also be applied to conservation tillage, but to a lesser extent.

There are some drawbacks to low- and no-till practices, however. Mulch can intercept a portion of the sprayed pesticide and interfere with surface coverage, thereby necessitating higher application rates for weed control. Shifts in weed population may occur that necessitate a change in herbicide selection and application methods. Plant pests also may become more of a problem in conservation tillage and no-till situations, necessitating more frequent applications. Foliage and mulch increases the surface roughness and exposed surface area, which increases the air turbulence at the surface. This results in an increase in the mass transfer rate from the surface due to the increased atmospheric turbulence above it and increases the vapor exchange rate, which increases volatilization.

4.2 TRANSPORT PROCESSES

LOCAL TRANSPORT

Once pesticides or related compounds have volatilized, they enter the surface boundary layer. The surface boundary layer has been described in terms of its potential temperature profile. Figure 4.1 shows that a large temperature gradient exists near the surface, with a nearly isothermal section forming the bulk of the layer, indicating that it is well mixed by turbulence. The slope of the potential temperature profile in the mixed layer may oscillate between positive and negative, but only small gradients occur because of a convective turbulence feedback mechanism.   This boundary layer forms over the surface of the earth and exhibits diurnal fluctuations in height that are dependent on surface properties such as roughness, temperature, and quantity and type of vegetation. The growth and height of the surface boundary layer is restricted by a capping inversion layer that is very stable.

The surface boundary layer performs a critical role in the vertical movement and horizontal distribution of airborne pesticides. The vertical movement of pollutants in the surface boundary layer is largely controlled by the prevailing atmospheric stability conditions (air temperature stratification). During the daytime, this boundary layer is usually unstably stratified, generally well mixed by mechanical and thermal turbulence, and typically extends several kilometers above the surface (Wyngaard, 1990). Any chemical released into the atmosphere under these conditions also will tend to become well mixed and dispersed throughout the surface boundary layer. At night, because of surface cooling, the boundary layer depth typically decreases to between a few tens to several hundred meters and is usually only slightly turbulent, quiescent, or very stable (Smith and Hunt, 1978). Chemicals released into a stably stratified atmosphere can be transported horizontally for long distances and generally undergo little mixing or dilution.

Local transport of pollutants (on the range of tens of kilometers) is confined to the environment surrounding the application area if they remain contained in the surface boundary layer (the lower troposphere). If they are rapidly transported to the mid- and upper troposphere (5 to 16 km), their residence times increase along with their range (Dickerson and others, 1987).

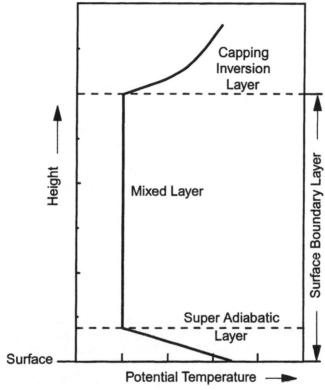

FIGURE 4.1. Profile of the surface boundary layer in terms of potential temperature with height (adapted from Tennekes, 1973).

REGIONAL AND LONG-RANGE TRANSPORT

Regional and long-range transport is defined as transport in the range of hundreds to thousands of kilometers from the point of application. Pollutant transport time into the free-moving troposphere above the surface boundary layer generally is on the order of a few weeks to months (Dickerson and others, 1987). Airborne pesticides can also move into the upper troposphere and stratosphere for more widespread regional and possible global distribution as a result of large-scale vertical perturbations that facilitate air mass movement out of the surface boundary layer. The transport time of an air parcel during large-scale vertical perturbations from the surface to a height of 10 km is on the order of hours, not months (Dickerson and others, 1987). Examples of large-scale vertical perturbations are:

- Large-scale convective instabilities such as "upsliding" at fronts where warm air masses are pushed over colder-heavier ones;
- Rotors and hydraulic jumps in mountainous regions that cause significant vertical mixing;

• Thunderstorm systems that can move air masses into the upper atmosphere; and
• The diurnal cycles of the surface boundary layer during which parcels of air may penetrate the capping inversion layer entrained in thermal plumes during the day, or which may remain aloft after the surface boundary layer height descends at night.

Once in the upper atmosphere, the global wind circulation patterns control long-range transport of airborne pollutants. The general global longitudinal circulation is a form of thermal convection driven by the difference in solar heating between the equatorial and polar regions. This general circulation is the result of a zonally symmetric overturning of the air mass in which the heated equatorial air rises and moves poleward where it cools, sinks, and moves equatorward again (Holton, 1979). The time-averaged motions of the atmosphere, where averages are taken over sufficiently long periods to remove the random variations associated with individual weather systems, but short enough to retain seasonal variations, show that trace species are lifted into the upper troposphere by the wind circulation cells (Figure 4.2). The air masses are carried poleward and descend in the subtropics, subpolar, and polar regions. These air masses are then carried back to the tropics in the lower atmosphere (Levy II, 1990).

In the Northern Hemisphere, the most intense atmospheric circulation occurs during the winter months when the temperature and pressure gradients are the steepest over the western perimeter of the North Atlantic Ocean (Whelpdale and Moody, 1990). Airborne pollutants from mid-latitude Eurasia and North America also are transported northward during the winter months (Barrie, 1986). This northward transport together with the lower ambient temperatures combine to increase the deposition rates of airborne pesticides into the Arctic and produce a warm-to-cold distillation effect (Goldberg, 1975; Cotham and Bidleman, 1991; Iwata and others, 1993).

Atmospheric concentrations of chlorinated pesticides such as HCH, HCB, DDTs, toxaphene, and chlordanes, have been observed in the Arctic, but the highest reported concentrations are generally α- and γ-HCH. This may indicate a vapor pressure dependence on global distribution profiles (Wania and others, 1992). Tanabe and others (1982) found that the highest air and seawater concentrations of DDTs and HCHs in global distribution correspond to the areas of the Hadley and Ferrel cells in the tropical and mid-latitude zones as did Tatsukawa and others (1990), but these areas are also located near the areas where these pesticides are used heavily.

Transport between hemispheres is limited due to the lifting of air parcels out of the surface boundary layer into the upper troposphere during storm events and the typical pole-eastward transport along usual storm tracts. Air masses do mix between the hemispheres, but this mixing time is on the order of 1 to 2 years (Czeplak and Junge, 1974; Chang and Penner, 1978; Ballschmiter and Wittlinger, 1991). Kurtz and Atlas (1990) and Iwata and others (1993) suggest that atmospheric transport of synthetic organic compounds is the major input pathway to most of the oceans of the world. Atlas and Schauffler (1990) suggest that the major sources of anthropogenic compounds in the Northern Hemisphere originate from the mid-latitudes.

4.3 REMOVAL PROCESSES

Once in the atmosphere, the residence time of a pesticide depends on how rapidly it is removed by deposition or chemical transformation. Both vapor and particulate-associated pesticides are removed from the atmosphere by closely related processes, but at very different rates. Atmospheric depositional processes can be classified into two categories, those involving

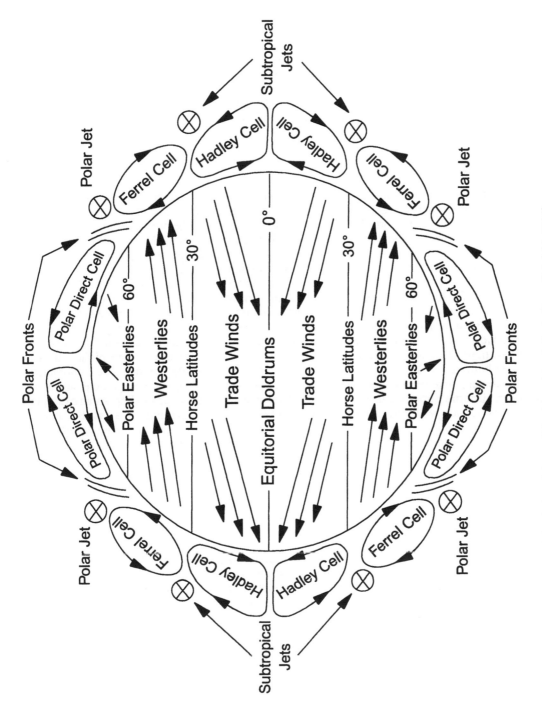

FIGURE 4.2. The general wind circulation of the earth's atmosphere (adapted from Seinfeld, 1986).

precipitation, called wet deposition, and those not involving precipitation, called dry deposition (Bidleman, 1988). Removal involving fog, mist, and dew lies somewhere between the wet and dry processes, but is more closely related to dry deposition. The effectiveness of the various removal processes depends on the physical and chemical characteristics of the particular compound, along with meteorological factors, and the underlying depositional surface characteristics. Either category of processes, however, involves both particle and gaseous transfer to the earth's surface. Figure 4.3 shows a generalized schematic of the distribution and deposition pathways. The partitioning of pesticide vapor into a raindrop, or sorption onto suspended particles, increases the effective size of the molecule as well as its atmospheric removal potential (Figure 4.4).

DRY DEPOSITION

In addition to the atmospheric introduction of pesticides sorbed to particles by wind erosion, pesticide vapors can sorb onto suspended particulate matter. The particulate matter may be relatively passive to the sorbed chemical or it may catalyze a chemical reaction or affect the photochemical process (Judeikis and Siegel, 1973; Behymer and Hites, 1985). Deposited particles and associated pesticides can be reintroduced to the atmosphere by rebound, reentrainment, or resuspension (Paw U, 1992; Wu and others, 1992).

Dry deposition of pesticides associated with particles includes gravitational settling, and turbulent transfer to a surface followed by inertial impaction, interception, or diffusion onto surfaces such as vegetation, soil, and water. The deposition rate is dependent on the size, surface area, and mass of the particle, and larger particles are greatly influenced by wind speed. Although larger particles usually weigh more than smaller ones and tend to settle out faster, most of the sorbed pesticide may be concentrated on the smaller particles because of their higher surface area-to-volume ratio (Bidleman and Christensen, 1979). As particle size decreases, buoyancy, viscous forces, and turbulence become more important in keeping the particle airborne. However, airborne particles can change size and become either larger or smaller. As an example, aerosols, which are relatively stable suspensions of solid or liquid particles in a gas (Finlayson-Pitts and Pitts, 1986), may coagulate to form larger droplets or particles, and large droplets and particles can break apart. Small particles also can react with atmospheric gases, be scavenged by precipitation, or act as condensation nuclei for water vapor.

The extent of vapor-particle partitioning can be estimated using equation 1 (Junge, 1977; Pankow, 1987),

$$\Phi = \frac{C_{ptcl}}{(C_{vapor} + C_{ptcl})} = \frac{c\Theta}{P_L{}^\circ + c\Theta} \tag{1}$$

where C_{ptcl} and C_{vapor} are the particle-phase and gas-phase atmospheric concentrations, respectively, Θ is the aerosol surface area, $P_L{}^\circ$ is the saturation, subcooled liquid-phase vapor pressure of the compound at the temperature of interest, and c is a constant that is dependent, in part, on the heat of vaporization, the heat of desorption, and the molecular weight of the compound. Dry deposition is a continuous, but slow process and is a function of the dry deposition velocity ($v_{d(z)}$), the deposition rate per unit area (F_d), and the airborne concentration ($C_{a(z)}$) (equation 2).

$$v_{d(z)} = -\frac{F_d}{C_{a(z)}} \tag{2}$$

The minus sign indicates a flux towards the surface. The deposition velocity and air concentration are both a function of height (z). Many variables influence the magnitude of $v_{d(z)}$ including particle size, meteorology, and surface properties. These variables introduce a great deal of uncertainty in $v_{d(z)}$ measurements and make it a difficult property to measure (Sehmel, 1980).

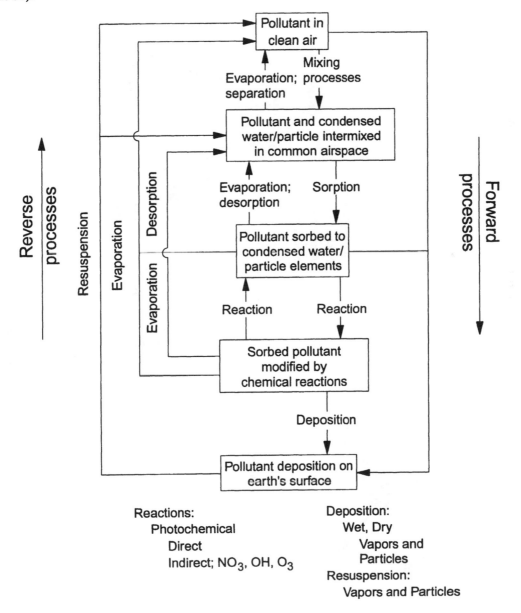

FIGURE 4.3. A simplified block diagram of gaseous and particulate pollution interconversion, and wet and dry deposition pathways (modified from Seinfeld, 1986).

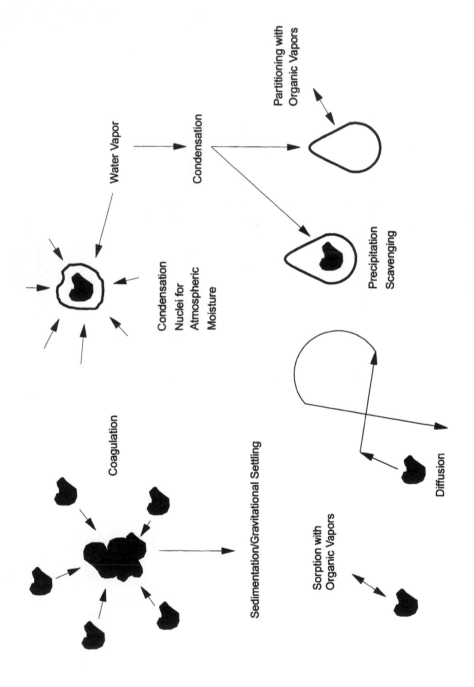

FIGURE 4.4. General diagram of the processes affecting airborne particulate matter.

WET DEPOSITION

Raindrops can act as a concentrating agent. They can concentrate cloud aerosols into droplets and scavenge vapor and particles as they fall through the atmosphere to the ground. One of the dominant mechanisms for removing persistent organic chemicals from the atmosphere is by rainout and washout (Ligocki and others, 1985a,b).   Rainout is the process where cloud droplets acquire contaminants within the cloud. Clouds form by the condensation of water vapor around nuclei such as particles or aerosols, both of which may contain organic contaminants. Washout is the process by which atmospheric contaminants are removed by rain below the clouds by the scavenging of particles and by the partitioning of organic vapors into the rain droplets or snowflakes as they fall to the earth's surface. Slinn and others (1978) estimated that a falling droplet will obtain equilibrium with a trace organic vapor within a distance of about 10 m, assuming the vapor concentration is constant throughout the path of the droplet. In reality, the falling droplet may encounter several different air masses in the fall to earth, and the vapor concentration of each air mass may affect the droplet concentration differently. For fine particles, precipitation scavenging is a more significant removal pathway relative to dry deposition because fine particles are airborne for a longer time than larger particles that have much higher depositional velocities (Glotfelty and Caro, 1975). Shaw (1989) observed that a 1-mm rainfall essentially cleansed the atmosphere of particulate matter. Others (Capel, 1991; Nations and Hallberg, 1992) have observed that the highest concentrations of pesticides in rain occur at the beginning of a rain event.

Total wet deposition (W) includes the deposition by rain of both vapor-phase and particle bound pesticides. The overall wet deposition can be approximated as the ratio of the total pesticide mass per volume rain ($C_{rain,\ total}$) to total pesticide mass per volume air ($C_{air,\ total}$) (equation 3).

$$W = \frac{C_{rain,\ total}}{C_{air,\ total}} \tag{3}$$

W is related to the washout ratios (vapor scavenging) of vapors (W_g), particles (W_p), and the fraction of pesticide associated with particulate matter (ϕ) (Pankow and others, 1984; Mackay and others, 1986). The partitioning of pesticide vapor into rain and cloud droplets (W_g) can be approximated by equation 4,

$$W_g = \frac{C_{rain,\ diss}}{C_{vapor}} = \frac{RT}{H} \tag{4}$$

where $C_{rain,\ diss.}$ is the dissolved-phase pesticide concentration in the droplet and C_{vapor} is the vapor-phase pesticide concentration. W_g also can be estimated as the reciprocal of the Henry's law value (H) where R and T are the universal gas law constant and the temperature (Kelvin), respectively.

Wet deposition of particles (W_{ptcl}) is defined in equation 5,

$$W_{ptcl} = \frac{C_{rain}}{C_{air,\ ptcl}} \tag{5}$$

where $C_{air,\ ptcl}$ is the particle bound concentration. W_{ptcl} is often determined experimentally from field sampling of nonvolatile species such as elemental carbon, ionic compounds, or trace metals (Cotham and Bidleman, 1991).

Total wet deposition (W) can be determined by measuring the total pesticide concentration in rain and the total pesticide concentration in air, or it can be estimated using W_g, W_p, and ϕ as in equation 6 (Mackay and others, 1986).

$$W = \frac{C_{rain, \, total}}{C_{air, \, total}} = W_g \, (1 - \phi) + W_p \phi \qquad (6)$$

CHEMICAL REACTIONS

Atmospheric chemical reactions are important as part of the removal process along with wet and dry deposition. They may result in products that are more toxic or more persistent, or both, than the original molecule. Photochemical reactions are the most important reaction type for airborne pesticides because these residues are totally exposed to sunlight. Reviews or articles on the photochemical reaction of herbicides (Crosby and Li, 1969; Crosby, 1976; Monger and Miller, 1988; Cessna and Muir, 1991; Kwok and others, 1992), insecticides (Turner and others, 1977; Woodrow and others, 1983; Chukwudebe and others, 1989), and fungicides (Schwack and Bourgeois, 1989) are presented elsewhere and general aspects are summarized below.

There are two processes by which an airborne pesticide can undergo a photochemical reaction: (1) By a direct process in which the pesticide absorbs sunlight directly and undergoes one or more of a variety of reactions; and (2) by an indirect process that involves reaction with photochemically generated oxidants such as ozone, hydroxyl radicals, ground-state atomic oxygen, or hydroperoxy radicals. These oxidants react with many organic compounds (Atkinson and Carter, 1984; Atkinson, 1989), including pesticides in the presence of light. The extent to which a compound can be photochemically degraded depends on characteristics particular to that compound. For direct reactions, a compound must absorb ultraviolet energy between 290 and 450 nm and its chemical structure must allow for breakdown or rearrangement. Generally, this means the compound must have unsaturated or aromatic bonds. For indirect reactions, the pesticide must react with the oxidant.

The source of airborne photoproducts is often difficult to ascertain. The photoproduct may form in the atmosphere by vapor phase reaction of the parent compound, or by photoreaction on a surface such as soil, foliage, or water followed by volatilization. Photolysis of the parent molecules also may occur when they are sorbed to airborne particulate matter or dissolved within the water droplets; however, sorption of pesticides to particles or dissolution into rain drops also may deactivate the pesticide to photochemical reactions. Photochemical reactions are more likely to occur within a rain droplet and other forms of atmospheric moisture because the pesticide concentration within the droplets may be higher than in the vapor phase. Surface films can form over the water droplet which can reduce the evaporation rate of the droplet as well as the air-water partitioning capability, thereby increasing the photochemical reaction time of the molecules (Gill and others, 1983).

The atmospheric photoreaction half-lives of certain classes of pesticides, such as organophosphates, may range from a few minutes to several hours (Woodrow and others, 1977; Woodrow and others, 1978; Klisenko and Pis'mennaya, 1979; Winer and Atkinson, 1990) or longer in some cases. Their transformation products may be less photoreactive and more long-lived. The main photoproduct of many organophosphorus pesticides is an oxygen analog that is usually more toxic than the parent, but in the case of parathion, the oxygen analog can be further transformed to the phenol and phosphates (Woodrow and others, 1983). Most oxidative reaction products are more polar than the parent compound. This suggests that they also will be more water soluble and more readily removed by wet-depositional processes or by air-water exchange.

CHAPTER 5

Analysis of Key Topics: Sources and Transport

The overview of the national distribution and trends of pesticides in the atmosphere, and the governing factors that affect their concentrations in the atmosphere, leaves many specific questions unanswered. Although some issues cannot be addressed on the basis of existing information, the most important topics deserve our best attempt. The following three chapters discuss in detail the key topics of sources and transport; phases, properties, and transformations; and environmental significance of pesticides in the atmosphere.

5.1 SEASONAL AND LOCAL USE PATTERNS

The overwhelming conclusions drawn from reviewing the studies listed in Tables 2.2, 2.3, and 2.4 are that the highest pesticide concentrations in air and rain are correlated to local use, and that locally high concentrations in rain and air are very seasonal. The highest concentrations usually occur in the spring and summer months, coinciding with application times and warmer temperatures. Insecticide concentrations in air and rain, however, are also high during the autumn and winter in correspondence to local use.

Nations and Hallberg (1992) detected clear areal and seasonal trends in herbicide detections in Iowa rain. The herbicides pendimethalin, EPTC, and propachlor were detected more frequently in the western part of the state where they were used more heavily than in the eastern part. The same trend was found for atrazine, alachlor, and cyanazine, which were used more in the northeastern part of the state. Pesticide detections in rain generally began in late April and continued through July or August. Concentrations were highest during April, May, and June, the months during which most of the pesticides are applied in Iowa. From August through November, pesticide detection frequency and concentrations in rain were much less, with no pesticide detections in December through March (see Figure 5.1). Pesticides were detected earliest in southern parts of Iowa where spring tillage and herbicide applications begin earlier than in northern parts. Goolsby and others (1994) found that triazine and acetanilide herbicide concentrations and detection frequency in rain were highest in the intense corn-growing areas of Iowa, Illinois, and Indiana (see Figure 3.10A). During the 2-year duration of the study, the concentrations and detection frequency increased in March, peaked during May and June, then decreased rapidly thereafter. Capel (1991) also found that concentrations of atrazine, cyanazine, and alachlor in rain peaked during the spring herbicide application season in Minnesota. Glotfelty and others (1990c), and Wu (1981) found the same spring-summer behavior for alachlor

131

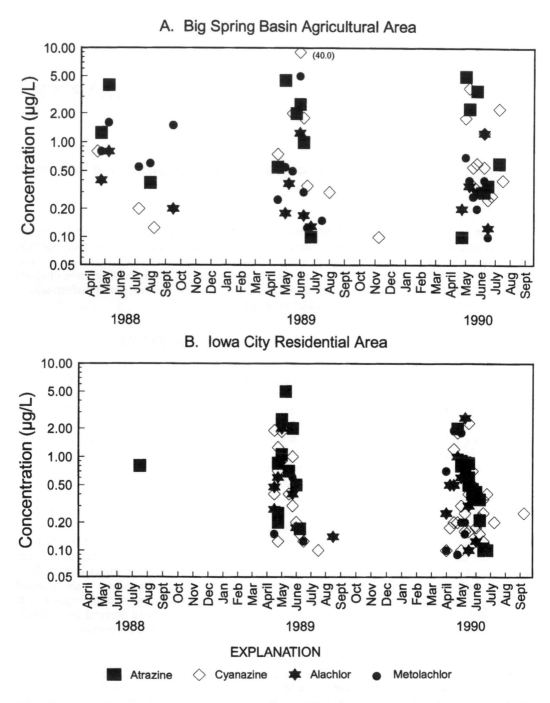

FIGURE 5.1. Detection frequency and concentrations for atrazine, cyanazine, alachlor, and metolachlor in Iowa rain. Data is for (A) an agricultural area and (B) an urban area between April 1988 and September 1990 (adapted from Nations and Hallberg, 1992).

(see Figure 5.2), metolachlor, atrazine, simazine, and toxaphene in rain and air at several sites throughout Maryland.

Trifluralin and triallate are two other high-use herbicides whose occurrence and concentrations in air have been found to correlate well with local use. Grover and others (1981, 1988a) found that the highest air concentrations of both herbicides occurred during May and June at several locations throughout Saskatchewan, Canada (see Figure 5.3). Air concentrations increased slightly during late October and November, which corresponded to the second application season. They observed that, during dry periods, the air concentrations decreased and that immediately after rain events the air concentration increased. Presumably this was due to the desorption of the herbicides from the remoistened soil, which resulted in an increase in volatilization.

Two large-scale, national studies that investigated the occurrence of pesticides in air at the same sampling locations for one or more years (Stanley and others, 1971; Kutz and others, 1976) found that the highest pesticide concentrations corresponded to local spraying and showed a seasonal periodicity. Arthur and others (1976) found that pesticide concentrations were highest during the summer months when use was highest. They found DEF, a cotton defoliant, only in

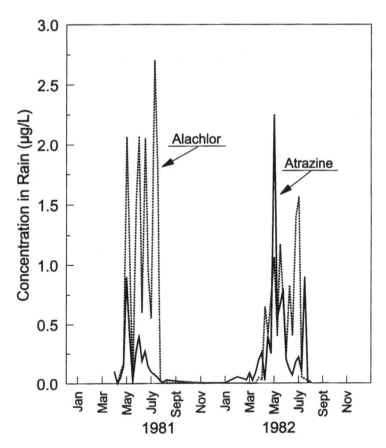

FIGURE 5.2. Seasonality and concentration range of alachlor and atrazine in Maryland rain in vicinity of Wye River (adapted from Glotfelty and others, 1990c).

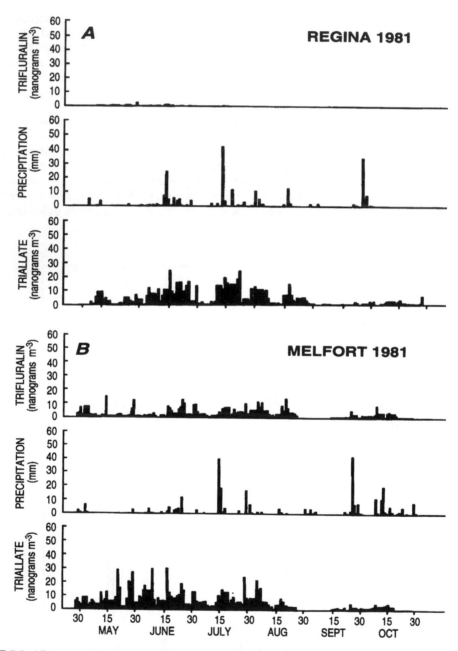

FIGURE 5.3. Histogram of triallate and trifluralin residues in air and the precipitation pattern during 1981 at (*A*) Regina and (*B*) Melfort, Saskatchewan (from Grover and others, 1988a), and triallate residues in air and the precipitation pattern during 1979 at (*C*) Indian Head, Saskatchewan (from Grover and others, 1981).

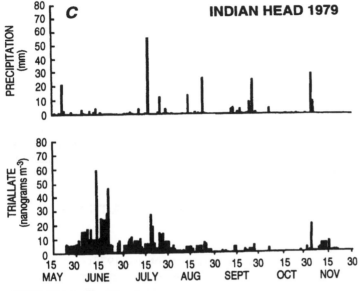

FIGURE 5.3.--*Continued*

September and October when it was used during cotton harvest season. Methyl parathion was detected in air from June through October, which are the usual application months in Mississippi. They also detected low methyl parathion concentrations during several winter months of 1974, but were unable to offer an explanation.

Harder and others (1980) reported that toxaphene concentrations were highest in midsummer rain over a South Carolina salt marsh and continued through September and October. The timing of these observations corresponded to the high agricultural use periods. During the winter when toxaphene use was very low, detectable concentrations in rain were infrequent. Rice and others (1986) also measured peak toxaphene air concentrations during September in Mississippi, Missouri, and Michigan. They continued sampling through mid-November and found that the air concentrations decreased during this time.

Pesticide occurrence in air, rain, and fog shows seasonal trends, with the highest concentrations corresponding to the growing seasons and local use that are not restricted to the spring and summer months. Shulters and others (1987) found that parathion, diazinon, malathion, 2,4-D, and γ-HCH concentrations in rain near Fresno, California, were highest between December and March, during the dormant spray season for fruit trees. Concentrations of chlorpyrifos, diazinon, methyl parathion, and methidathion, which are also used as dormant sprays, were high in winter fog near the same area (Glotfelty and others, 1987; Seiber and others, 1993). Pesticides also have been detected during periods before and after the use and growing season; however, determining their sources has proven difficult. These nonseasonal occurrences could be due to volatilization or wind erosion of previously applied material, or both, or the result of long-range transport from areas whose growing season started earlier or later (Glotfelty and others, 1990c; Wu, 1981).

The seasonality of occurrence in air and precipitation seems to be true for those pesticides in current use as well as those that are in limited use or no longer used in the United States and Canada, such as the organochlorine pesticides DDT, dieldrin, and toxaphene. Brun and others (1991) found that concentrations of α- and γ-HCH in precipitation were highest during the spring and autumn months at three sites in Atlantic Canada. Kutz and others (1976) found the highest detection frequency and air concentrations throughout the United States occurred from May through September, as did Hoff and others (1992) for Egbert, Ontario. Hoff and others (1992) also sampled the air for various other halogenated pesticides such as DDTs, chlordanes, toxaphene, dieldrin, endosulfan, trifluralin, endrin, and heptachlor and found that all of these compounds, whether in current use or not, exhibited maximum air concentrations during the spring and summer months (see Figure 5.4). Apparently, the source of airborne organochlorine compounds that are no longer used in the United States is the volatilization of residues remaining in the treated fields (Seiber and others, 1979; Tanabe and others, 1982; Bidleman and others, 1988). Another source is atmospheric transport into the United States from countries such as Central and South America, Eastern Europe, and Asia where these pesticides are still extensively used (Rapaport and others, 1985; Bidleman and others, 1988).

5.2 EFFECTS OF AGRICULTURAL MANAGEMENT PRACTICES

Agricultural management practices include pesticide application methods and formulations, irrigation methods, and tillage practices. Maybank and others (1978) compared the amount of drift of aqueous solutions of a 2,4-D ester applied by ground-rig and aerial pesticide application systems. The drift during the ground-rig applications ranged from less than 0.5 to 8 percent of the nominal application and was dependent on the nozzle type, hydraulic pressure, and windspeed. The drift from aircraft applications ranged from 1 to 31 percent. Frost and Ware (1970) compared the drift from several types of ground applications to aerial applications. They found that the ground-rig sprayer applications had 4 to 5 times less drift than aerial applications and 4 to 10 times less drift than ground mist-blower applications. They also found that aerial application drift was up to 2 times less than that from ground mist-blower applications. Aerial spray drift can be reduced by flying closer to the ground, but when the aircraft is too close, the wing-tip vortices cause the spray cloud in the wake of the aircraft to actually rise (Lawson and Uk, 1979), which enhances the drift potential. Controlling drift from mist-blowers is difficult because they, generally, produce a smaller droplet size and are propelled into relatively calm air at velocities in excess of 145 kilometers per hour (90 miles per hour) (Ware and others, 1969).

The physical placement of the pesticide also has been shown to affect post-application volatilization. Bardsley and others (1968) found in a laboratory experiment that placing trifluralin 1.27 cm below the soil surface reduced the vapor loss by a factor of about 25 times that of surface-applied losses. Another laboratory study (Spencer and Cliath, 1974) found that the vapor loss rate of trifluralin incorporated into the top 10 cm of soil was 51.7 g/ha/d for the first 24 hours, while the surface-applied losses were 4,000 g/ha/d. Actual field measurements showed that triallate and trifluralin incorporated to a depth of 5 cm volatilized at a maximum rate of 4 and 3 g/ha/d respectively, during the first 4 to 6 hours after application (Grover and others, 1988b). The volatilization rate of surface-applied triallate and trifluralin was 70 and 54 g/ha/d, respectively, for the same period of time (Majewski and others, 1993). The same type of results was reported for the insecticides fenitrothion and deltamethrin when they were applied to the surface or injected into water (Maguire, 1991). Pesticides also can be applied through irrigation water. Cliath and others (1980) found that for EPTC, 74 percent of the applied amount was lost by volatilization in the first 52 hours. They concluded that this application technique was very inefficient for EPTC.

Seiber and others (1989) found a qualitative correlation between daily measured air concentrations and local use for methyl parathion, molinate, and thiobencarb in a rice-growing area of northern California. This relation was strongest for methyl parathion. All three pesticides

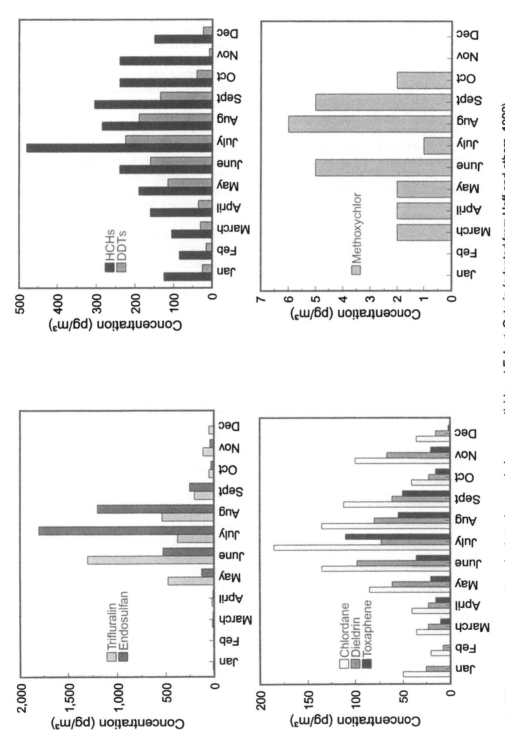

FIGURE 5.4. Air concentrations of selected organohalogen pesticides at Egbert, Ontario (adapted from Hoff and others, 1992).

were applied by aircraft, but methyl parathion was applied as a water-based emulsifiable spray and the other two were applied as granular formulations. The closer correlation of air concentrations to use for methyl parathion was attributed to drift of the vapor and fine aerosol component of the liquid spray during application. There was very little measured drift associated with granular applications. The primary source of molinate air concentration was post-application volatilization (Seiber and McChesney, 1986), which occurred continuously after the application. The rate of molinate post-application volatilization was influenced, in part, by its vapor pressure, which is about 300 times that of methyl parathion.

Turner and others (1978) investigated the effects that different carrier formulations had on pesticide drift and volatilization during and post-application for chloropropham. They found that a microencapsulated formulation of the herbicide reduced both. The emulsifiable formulation had about 5 percent drift loss of the nominal application, whereas the microencapsulated formulation had less than 1 percent drift loss. The effect was largest in the post-application volatilization where the emulsifiable formulation volatilized at five times the rate of the encapsulated formulation.

Wienhold and others (1993) looked at the effects that starch encapsulation, liquid commercial formulations, and temperature had on the volatilization of atrazine and alachlor using agroecosystem chambers. They found that for atrazine the volatilization rate of the commercial formulation was nearly 12 times higher than the starch-encapsulated form at 15°C and nearly 5 times greater at 25 and 35°C. Alachlor showed the opposite behavior, with the starch-encapsulated formulation volatilizing 1.3 times faster than the commercial form at 15°C and 3.3 times faster at 25 and 35°C. This difference in herbicide behavior was attributed to their chemical properties. These results show that one management practice cannot be used across the board for all pesticides. The physical and chemical properties of the pesticide dictate the best use methodology.

Tillage practices such as conventional-, low-, and no-till, and the potential effects each has on pesticide inputs into the lower atmosphere have been discussed in Chapter 4, Section 4.1. An actual comparison of the effects that different tillage practices have on pesticide volatilization was reported by Whang and others (1993). They compared the volatilization losses of fonofos, chlorpyrifos, and atrazine from a conventional- and a no-till field. The results showed that the no-till field had 26-day cumulative volatilization losses for fonofos, chlorpyrifos, and atrazine that were 2.3, 4.1, and 1.3 times greater than those of the conventionally tilled field, respectively. They speculated that the no-till field volatility losses were greater than the conventionally tilled field because the mulch provided a greater surface area for contact between the pesticide residue and air.

Nations and Hallberg (1992) detected a greater variety and higher concentrations of herbicides than insecticides in Iowa rain. This may have been due to the greater use of herbicides in Iowa, but it is also quite possible that the application method played an important part. Herbicides are usually sprayed on the surface in liquid formulations, while insecticides are often applied as granular formulations and incorporated into the soil. This may explain why chlorpyrifos and terbufos, which are both heavily used in Iowa agriculture, were not detected in any of their rain samples.

The contribution of pesticide-bound soil particles to the total atmospheric burden is largely unknown. Glotfelty and others (1989) found that the post-application volatilization fluxes of a wettable powder (WP) formulation of atrazine and simazine exhibited wind erosion characteristics when measured over dry soil, but concluded that the amount of pesticide entering the atmosphere on wind-eroded WP formulation particles was small in comparison to the amount injected by true molecular volatilization for those pesticides with appreciable vapor pressures. Ross and others (1990) found an increasing percentage of the total downwind air concentration of dacthal, which was applied to an experimental field as a WP formulation, associated with particulate matter. This coincided with the drying of the applied soil surface. Greater than 30 percent of the off-site air concentration was retained on glass fiber filters, which was attributed to windblown dust. These results are consistent with those reported for dacthal by Glotfelty (1981). Menges (1964) found that the efficacy of five herbicides broadcast-sprayed to bare soil decreased by about 40 percent following a windstorm, which caused considerable erosion of the soil surface. He also found that when herbicides were applied to the soil in an established crop bed followed by moderate winds, the weed control was reduced but crop damage increased.

Very little work has been done on the resuspension of pesticides deposited to surfaces. The environmental influences on particle resuspension rates include windspeed, particle properties, relative humidity, surface properties, and exposure duration (Nicholson, 1988a; Wu and others, 1992). These are just for particulate matter with no chemical reactivity or vapor pressure. Pesticide resuspension, whether in vapor or particle form, depends on the distribution behavior between the vapor-particle and the vapor-aqueous phases as well as the surface characteristics. Particle resuspension has primarily been studied in arid and semiarid regions of the world (Sehmel, 1980; Nicholson, 1988b) and has dealt with erosion of deserts and agricultural areas (Chepil, 1945; Gillette, 1983). Wu and others (1992) showed a tremendous variability in the measured resuspension rates and Paw U (1992) showed that the rebounding and reentrainment of particles can decrease the overall net deposition to zero in some cases.

5.3 URBAN AREAS

Urban pesticide use is not as well documented or as studied as is agricultural pesticide use. Urban pesticide use includes individual consumer and professional applicators in home and industrial settings such as turf management in lawn and landscape care, golf courses, parks, cemeteries, roadways, railroads, and pipeline (Hodge, 1993). State and local municipalities use pesticides in the maintenance of parks, recreational areas, and right-of-ways. Pesticides are also used in large-scale control of pests, such as the mosquito, the Japanese beetle, the gypsy moth, and the Mediterranean fruit fly. In home use, the pesticide application rates are specified on the product, but the actual application rates are unregulated and no training is required. The professional applicators, however, commonly require training and licensing (Hodge, 1993). In agriculture, the application of pesticides often occurs in one large application, usually within a 2 to 3 week period around planting. Home lawn care and garden chemical use are often split into 3 to 5 small applications throughout the spring and summer months (Gold and Groffman, 1993). The results of a USEPA national home and garden pesticide use survey (Whitmore and others, 1992, 1993) for 113 ingredients commonly used around the home (Table 3.1) are reported as the number of products and the number of outside applications, rather than actual amounts in pounds applied, so it was difficult to make any meaningful comparisons to agricultural use.

Few studies have investigated pesticide concentrations in urban atmospheres, or compared urban pesticide use to agricultural pesticide use. Bevenue and others (1972) found the highest levels of p,p'-DDT, dieldrin, and lindane in rain at Honolulu, Hawaii, a large, crowded mix of residential, commercial, and industrial establishments. They detected lower concentrations in three other, primarily residential, areas of the island. Que Hee and others (1975) concluded that spraying in urban areas could sometimes cause more pollution than spraying in rural areas in their study of 2,4-D air concentrations in central Saskatchewan, Canada. Grover and others (1976), however, pointed out that this study did not correlate the high air concentrations with wind direction and that it did not rule out the possibilities of accidental spills near the sampling sites.

Nations and Hallberg (1992) detected atrazine, alachlor, and cyanazine with the same frequency in rural and urban sampling sites, but the concentrations were slightly higher at the rural sites (Figure 5.1). The only organophosphorus insecticides they detected were malathion, methyl parathion, dimethoate, and fonofos and only at the two urban sites, presumably because of their high lawn and garden use. The insecticides malathion and dimethoate were not used to any appreciable extent in Iowa agriculture in 1988 (Gianessi and Puffer, 1992b), but fonofos was ranked third (664,613 lb a.i./yr on corn), behind terbufos (1,520,743 lb a.i./yr on corn) and chlorpyrifos (1,395,794 lb a.i./yr on corn, sweet corn, and alfalfa). Methyl parathion was used to a lesser extent (69,630 lb a.i./yr on corn and apples). Terbufos and chlorpyrifos were not detected in any sample. Considering the very high use of these pesticides in Iowa, this finding was unexpected and not explained. One of the main findings of this study was that, while each sampling site within a specific area of Iowa contained the same suite of detected compounds, those sites closest to the sites of actual pesticide use contained the highest concentrations.

Of the three national scale studies done in the mid-1950's to early 1970's, Tabor (1965) investigated the occurrence of various pesticides (aldrin, chlordane, DDT, malathion, toxaphene) in air at various urban locations near agricultural areas and in communities with active insect control programs. He found substantial amounts of those pesticides used in or near each location in the air at all sites. Tabor also found that the concentrations in urban areas with active insect control programs were significantly higher compared to those near agricultural areas but concluded that the resultant human exposures were more intermittent and of shorter duration. Stanley and others (1971) sampled air at four urban and five rural locations and found the highest pesticide concentrations in agricultural areas of the south (DDTs, toxaphene, methyl parathion) and in one urban area (DDTs, HCHs, 2,4-D) in the west. This urban area, Salt Lake City, Utah, was reported to have considerable mosquito control activity during the sampling periods. Kutz and others (1976) sampled air at three urban locations: Miami, Florida; Jackson, Mississippi; and Fort Collins, Colorado. They found that both the Miami and Jackson samples contained higher concentrations and a greater variety of pesticides than did the Fort Collins samples. In their 16-state study in 1970-72, which targeted areas of high probability of detection, they found an average of 17 different pesticides in each of 16 states, with only 11 in Miami and Jackson, and 5 in Fort Collins.

5.4 RELATIVE IMPORTANCE OF LOCAL, REGIONAL, AND LONG-RANGE TRANSPORT

The distance that airborne pesticides are transported depends upon the removal rates (dry and wet deposition and chemical reactions). The highest atmospheric concentrations usually are associated with locally used pesticides and are seasonal in nature. During these high-use periods, any regional and long-range inputs are usually insignificant in comparison and lost in the background.

Examples of local atmospheric movement (tens of kilometers and mainly confined to the area surrounding the application areas) of pesticides are best described by spray drift during application and post-application volatilization followed by off-site drift. Spray drift has been recognized for its potential for nontarget crop damage since the mid-1940's (Daines, 1952). Seasonal high atmospheric concentrations of locally used pesticides have been shown to cause illegal residues on nontarget crops (California Department of Food and Agriculture, 1984-1986; Turner and others, 1989; Ross and others, 1990) as well as crop damage (Daines, 1952; Reisinger and Robinson, 1976). Research on various crops has studied the effects of low level exposure to various herbicides in the attempt to quantify actual crop yield losses (Hurst, 1982; Jacoby and others, 1990; Snipes and others, 1991, 1992). Drift and deposition during application is not the only time drift occurs. As explained in Chapter 4, many pesticides volatilize directly into the atmosphere from the target surface. Their volatilization rates can be determined, and their downwind air concentrations and deposition rates can be measured or estimated using various computer models. Rain can enhance the deposition rate by "washing out" the pesticide as the raindrops fall through the atmosphere. Pesticides have been found in fog (Glotfelty and others, 1987), which has been implicated in the deposition of airborne pesticides onto nontarget crops (Turner and others, 1989). Dew and frost may be another mechanism by which airborne pollutants are deposited to the surface (Foster and others, 1990).

Pesticide occurrence in air, rain, fog, and snow shows clear seasonal trends, with concentrations being greatest during the local use and growing season. Pesticides have been detected before and after the use and growing season, however, and determining their sources has proven difficult. These pre- and post-season residues could be due to volatilization and wind erosion of previously applied material. They also could be the result of long-range transport from areas whose growing season started earlier.

Several examples of regional pesticide movement are reported in the literature. Toxaphene and other organochlorine pesticides such as DDTs, chlordanes, and HCHs have been detected in the air, rain, snow, surface water, and soil in the Great Lakes drainage basin (Bidleman and Olney, 1974; Eisenreich and others, 1981; Rapaport and others, 1985; Rapaport and Eisenreich, 1986; Rice and others, 1986; Rapaport and Eisenreich, 1988; Bidleman and others, 1988; McConnell and others, 1993). Toxaphene was extensively used in cotton and soybean production in the southern United States between 1972 and 1982 before it was banned, but its use in the Great Lakes region was always very limited. It is still used in Mexico, Czechoslovakia, Poland, Hungary, and other countries around the world (U.N. Food and Agriculture Organization, 1978-87). Toxaphene and other pesticides are transported into the Great Lakes region by southerly winds from the Gulf of Mexico that flow in a northeasterly direction. In fact, a concentration gradient that increases from the north to the south along this air transportation corridor was measured by Rice and others (1986). Rapaport and Eisenreich (1986) showed a good correlation between toxaphene concentrations in air at various sites in eastern North America and the dominant air circulation pathways from major source areas to the south. The major environmental dissipative route for toxaphene is volatilization into the atmosphere (Seiber and others, 1979), and the current atmospheric concentrations are partly due to the volatilization of persistent soil residues (Bidleman and others, 1988; Bidleman and others, 1989).

Information on long-range and regional atmospheric transport (hundreds to thousands of kilometers) of pesticides is limited, but there is increasing recognition of this important area and the information database is growing. Risebrough (1990) described the airborne movement of pesticides from their point of application as a global gas-chromatographic system where pesticide molecules move many times between the vapor-soil-water-vegetation phases in maintaining an equilibrium of chemical potential (fugacity) between these phases. He stated that the movement of these chemicals from point A to point B does not adequately describe their environmental transport behavior. Once deposited on the earth's surface, the pesticide can revolatilize, or become reentrained into the atmosphere and be transported and deposited downwind again and again, until it is finally degraded or becomes distributed world-wide.

Most pesticides applied in the tropical areas rapidly volatilize into the atmosphere due to the high temperature climate of the area (Tanabe and others, 1982), and the use of organochlorine pesticides such as HCH and DDT remains high in some areas of the world. The concentration distribution of these compounds in the air and water of the world's oceans has shifted from the mid-latitude oceans of the Northern Hemisphere to the low-latitude areas (Bidleman and Leonard, 1982; Iwata and others, 1993). This reflects a shift in use from the developed countries of the United States, Europe, and Japan in the 1960's, through the 1980's, to the developing countries, primarily the tropical Asian countries (Goldberg, 1975; Agarwal and others, 1987; Kaushik and others, 1987; Tatsukawa and others, 1990; Iwata and others, 1993).

The large-scale circulation patterns (Hadley cells, Ferrel cells, Polar Direct cells) combine to move tropical air masses and the associated pesticides northward in the Northern Hemisphere and southward in the Southern Hemisphere (Figure 4.2). Several studies have inferred long-range transport of pesticides into their study area by the presence of pesticides in the rain before any local use began (Wu, 1981; Nations and Hallberg, 1992). Glotfelty and others (1990c) found low concentrations of atrazine and simazine in Maryland rain before the corn-planting season began and concluded that their presence was due to movement up from the southern coastal states whose planting season had already begun.

This same type of phenomenon has been observed in Europe. Air masses that have passed northward over eastern Europe into the Scandinavian countries have been shown to deposit organochlorine pesticides such as toxaphene, DDT, HCHs, HCB, and chlordane, as well as other compounds by wet and dry deposition processes in areas where their use is very low or nonexistent (Björseth and Lunde, 1979; Schrimpff, 1984; Barrie and Schemenauer, 1986; Bidleman and others, 1987; Pacyna and Oehme, 1988; Bidleman and others, 1989; Shaw, 1989). Amundsen and others (1992) reported that the same phenomenon occurred with the transport of trace elements from eastern Europe into southern Norway. Tarrason and Iversen (1992) concluded that North America contributes significant amounts of sulfur to western Europe, mainly through wet deposition; Connors and others (1989) followed an air mass containing elevated carbon-monoxide concentrations from Europe into the Middle East; and Whittlestone and others (1992) measured high radon levels over Hawaii that were determined to originate in Asia. Swap and others (1992) have found that parts of the Amazon rain forest are dependent on soil dust that originates in the Sahara/Sahel region of West Africa. This could be a primary mechanism for transport of particle-bound pesticides used in the West Africa region into the Amazon basin.

The Arctic and Antarctic are two areas where pesticides are not used, yet they are found in the air, snow, people, and animals there (Hargrave and others, 1988; Patton and others, 1989; Bidleman and others, 1990; Gregor, 1990; Muir and others, 1990). Eurasia (United Kingdom, Europe, the former Soviet Union) seems to be the most important source for the pollution-derived aerosol component of the Arctic haze (Rahn, 1981; Patton and others, 1991) as well as organochlorine pesticides (Bidleman and others, 1989; Patton and others, 1989). The Eurasian air mass moves east and northeast on the mid-latitude westerlies along the Pacific Polar front, which is a major storm track into the Arctic. Low concentrations of organochlorine pesticides are transferred through the food chain, which can reach high levels in mammals, especially in the Arctic regions. Addison and Zinck (1986) found that DDT concentration in the Arctic Ringed Seal did not decrease significantly between 1969 and 1981, while PCB did. This suggests that long-range transport deposits DDT into the region from those areas in eastern Europe where its use continues, but where PCB use has decreased.

The North American continent does not directly contribute much pollution to the Arctic because the continental wind flow patterns are generally away from the north pole. North America is, however, a major source of organochlorine compounds into the atmosphere. Because of their long atmospheric residence times, organochlorine pesticides can become well mixed throughout the troposphere, and long-range atmospheric transport of pesticides that are no longer used in the United States, as well as those in current use, can be a primary source of contamination to pristine areas where little or no pesticides are used. The environmental significance as well as the human health effects of continuous exposure to low concentrations of organochlorine and the variety of other pesticides detected in the atmosphere is largely unknown.

5.5 EFFECTS OF CLIMATE

Climate dictates which crops can be grown in different areas of the country, the actual growing seasons, crop rotation patterns, and which pests are present. For example, in the Midwest, corn and soybeans are the predominant crops grown during spring and summer, while in certain parts of California and the southeastern United States, a variety of crops are grown throughout the year. Stanley and others (1971) reported that pesticide air concentrations generally were highest during the summer months, when pesticide use was the highest, at their nine sampling locations throughout the United States, except for the Florida site, which had agricultural activities occurring throughout the year and corresponding high pesticide concentrations in air. Nations and Hallberg (1992) reported that pesticide detections in rain began earlier in the southern part of Iowa where the planting season began earlier than in the northern part of the state. Glotfelty and others (1990c) detected atrazine in Maryland precipitation before the planting season began and speculated that atrazine was being transported northward from the Gulf Coast states where corn planting began 1 to 2 months earlier.

The type of crop usually dictates which pesticide will be used as each pesticide is registered for use only on specific crops and the highest pesticide use generally occurs during the spring and summer months in most parts of the country. Herbicides used in corn and soybean production (primarily atrazine, alachlor, cyanazine, and metolachlor) have been detected most frequently and at the highest concentrations in rain in the Midwest (Nations and Hallberg, 1992; Goolsby and others, 1994). In other parts of the country where corn and soybeans are not predominate crops, these herbicides are detected less frequently and at lower concentrations, if at all (Glotfelty and others, 1987; Richards and others, 1987). Even though the climate of an area dictates which crops can be grown during certain times of the year, fluctuations in weather often require adjustments in the timing and amount of pesticide applications. As an example, unseasonably cool temperatures can retard plant growth rates, which could lead to early season pest problems that may necessitate earlier, different, or more frequent pesticide applications. Droughts and heavy rains also have special pest problems associated with them.

Climate also dictates the behavior of airborne pesticides. The ambient temperature plays an important role in the vapor-particle partitioning behavior (Yamasaki and others, 1982; Bidleman and others, 1986; Pankow, 1987, 1991, 1992; Pankow and Bidleman, 1991). Larsson and Okla (1989) found that PCB, DDT, and DDE concentrations in Swedish air were positively correlated with temperature. They found the summer air concentrations primarily were associated with the gas phase whereas the winter air concentrations primarily were associated with the particle phase. Patton and others (1991) found that in the Arctic at -28°C, compounds with saturation liquid-phase vapor pressures of $P_L° \geq 10^{-3}$ at -28°C, such as pentachlorobenzene, were almost entirely in the vapor phase. Those compounds with $10^{-6} \leq P_L° \leq 10^{-3}$ at -28°C, such as α- and γ-HCH, dieldrin, cis- and trans-chlordane, o,p'-DDE, and p,p'-DDE, were distributed between the particle and the vapor phase, and those compounds with $P_L° \leq 10^{-6}$ at -28°C, such as p,p'-DDT, and five or more ring PAHs, were almost entirely associated with the particle phase. These results are in agreement with predictions from equation 1; however, uncertainties exist from the extrapolation of experimental vapor pressures, generally determined at room temperature, to -28°C. In general, the less volatile a compound is, the more it will associate with suspended particulate matter, and the ambient temperature directly influences vapor pressure and, therefore, the volatility potential of pesticides and other organic compounds.

Wind direction also can have an effect on pesticide concentrations in air. Burgoyne and Hites (1993) found a positive correlation between the atmospheric concentrations of endosulfan, ambient air temperature, and an easterly wind direction in Bloomington, Indiana. They found the highest concentrations between April and August (see Figure 5.5). Lane and others (1992) also reported that atmospheric concentrations of α- and γ-HCH in Ontario, Canada, depend on temperature and wind direction.

Photochemical reactions are important mechanisms for removing pesticides from the atmosphere and are dependent on the duration, degree of dispersion, and intensity of solar energy reaching the lower atmosphere and the surface of the earth. The intensity increases with the clarity of the atmosphere, time of day, and altitude. As an example, the average daily solar energy received during May in Alaska or Washington (564 and 620 gram-calorie per square centimeter, respectively) can be more than in southern California or northern Georgia (482 gram-calorie per square centimeter each) (Crosby and Li, 1969). The time of day, as well as the time of year, can influence photochemical reaction rates. Turner and others (1977) reported that photodieldrin formation and volatilization peaked between noon and 2:00 p.m. during a July field experiment in Maryland, and Woodrow and others (1977, 1978) found that the half-life for parathion was on the order of 2 minutes at noon in June in California and about 131 minutes after sunset. Woodrow and others (1978) also found that the atmospheric half-life of trifluralin was 21 minutes during optimum sunlight conditions at noon in August and increased to 3 hours in October in California. Glotfelty and others (1990a) speculated that the oxygen analogs of chlorpyrifos, diazinon, methidathion, and parathion found in California fog water were primarily formed in the atmosphere during the daylight and partitioned into the fog as it formed at night.

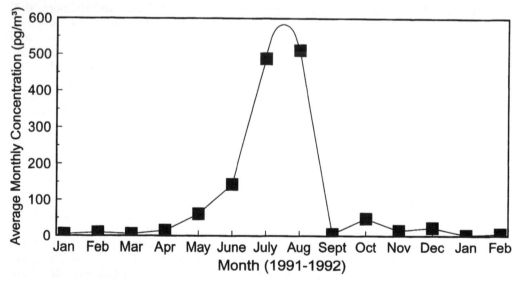

FIGURE 5.5.   Average monthly air concentrations of endosulfan between January 1991 and February 1992, Indianapolis, Indiana (adapted from Burgoyne and Hites, 1993).

CHAPTER 6

Analysis of Key Topics: Phases, Properties, and Transformations

Numerous mechanisms can deliver pesticides to the atmosphere. Once in the atmosphere, pesticides are distributed among the aqueous, particle, and vapor phases. This distribution, along with transformation reactions, strongly affects the behavior, transport, and ultimate fate of airborne pesticides. Numerous mechanisms also deliver pesticides back to the surface of the earth. These include wet deposition, such as rain, snow, and fog, and dry deposition of vapor-phase and particle-bound pesticides.

6.1 INFLUENCE OF CHEMICAL AND PHYSICAL PROPERTIES

Mechanisms that deliver pesticides to the atmosphere that are chemical in nature, such as volatilization from soil and volatilization from water, can be described by structure-activity relations and predictions of their importance made for any given chemical. Based on volatilization predictions, the pesticides with very high vapor pressures would be the ones most likely to be present in the atmosphere if they are not rapidly transformed. There are also meteorological entrance mechanisms that are physical in nature, such as wind erosion and drift during and after application. Wind speed and ambient temperature also affect the volatilization rate of a pesticide. Predictions of the importance of these types of mechanisms are not always straightforward.

In Figure 6.1, it can be seen that the pesticides detected in the atmosphere (denoted by ◊) are interspersed among those pesticides that have not been measured or detected in the atmosphere (denoted by ◆), even though the chemical properties are very similar. There are several reasons that help to explain why a particular pesticide has not been detected. These may include low use, short atmospheric residence time (considering deposition and transformation), the timing of the measurement relative to the timing of use, the predominant atmospheric phase in which it will accumulate relative to the phase being sampled and, perhaps most important, whether or not the pesticide has been looked for in the atmosphere. From the evidence in the literature, essentially all pesticides that have been targeted for analysis in atmospheric studies have been detected in at least one atmospheric matrix. That is to say, many of the other pesticides from Figure 6.1 that have not been measured in the atmosphere probably are present at detectable concentrations at some time of the year in some locations, but just have not been targeted for analysis. Some of these may exist in the atmosphere for only short periods of time or over short distances due to their strong affinity for atmospheric particles (very low vapor pressures) or very fast transformation kinetics. For those pesticides with fast transformation kinetics, measurements of transformation products may be very important in determining their fate in the atmosphere.

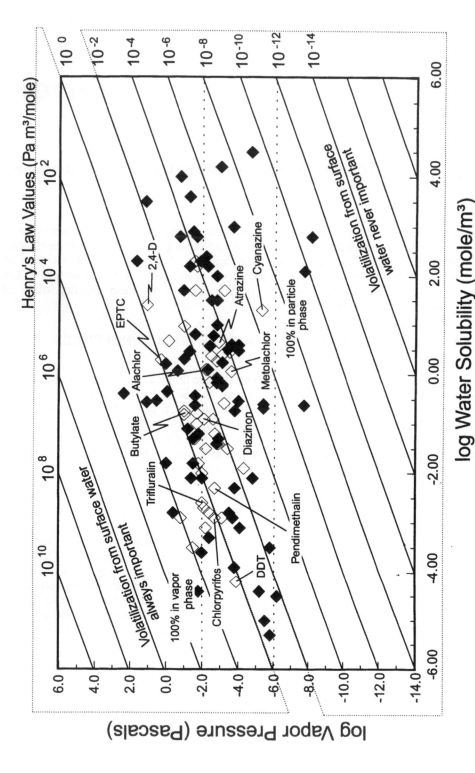

FIGURE 6.1. Relation between vapor pressure and water solubility for various pesticides from Table 6.1. The open symbols (◇) represent those pesticides detected in the atmosphere. Several currently high-use pesticides and DDT are identified.

6.2 PHASE DISTRIBUTION AND TRANSFORMATION REACTIONS

The distribution of pesticides among aqueous, particle, and vapor phases strongly affects their behavior, transport, and ultimate fate. Regardless of the mechanism by which a pesticide enters the atmosphere, it will distribute among all three phases as equilibrium conditions are approached. The equilibrium condition for a particular pesticide in the atmosphere is dependent on the properties of that chemical, including water solubility and vapor pressure and the characteristics of the atmosphere, including the temperature, moisture content, and nature and concentrations of particulate matter.

The phase a pesticide is associated with strongly affects its removal potential from the atmosphere by wet and dry depositional processes. Pesticides with low vapor pressures of less than 10^{-6} Pa (Figure 6.1) will exist primarily in the particle phase in a normal rural atmosphere (Bidleman, 1988) and be most prone to removal from the atmosphere by dry deposition and rain scavenging of particles. Pesticides with high vapor pressures (greater than 10 Pa) will primarily exist in the vapor phase, and dry deposition of particles will not be as important in their atmospheric removal. Table 6.1 lists the water solubility (S) and vapor pressures (VP) for selected pesticides in current use, as well as for several organochlorine pesticides that are in limited use or no longer used in the United States. Pesticides with high vapor pressures are the least efficiently removed from the atmosphere by wet depositional processes, and pesticides with high water solubilities are the least efficiently removed from the atmosphere by dry depositional processes. Pesticides with these characteristics will tend to accumulate in the atmosphere until they are removed or altered by transformation reactions. The same generalities can be made for the Henry's law constant (H) values (Table 6.1). For pesticides with H values less than about 1 Pa-m^3/mole, removal by vapor-water transfer into raindrops is an important control on their atmospheric concentration. For pesticides with H values greater than this, removal by falling raindrops is less important. As H increases, the importance of removal of chemicals from the atmosphere by forms of condensed water diminishes. Pesticides with high vapor pressures and high water solubilities are the least efficiently removed from the atmosphere by wet and dry depositional processes, respectively, and will tend to accumulate in the atmosphere until they are removed or altered by transformation reactions.

It should be noted that the values listed in Table 6.1 are taken from several compilations of physical and chemical properties of pesticides. In order to facilitate the calculation of Henry's law values, the vapor pressure and water solubility data from these references have been converted to the subcooled liquid form using the fugacity ratio, F (Suntio and others, 1988). Subcooled liquid values are the most accepted, environmentally accurate way of expressing vapor pressure and water solubility data. This conversion requires that the entropy of fusion (ΔS) be known, which for most pesticides is not available. For those pesticides, ΔS is estimated as 56 Joule per mole Kelvin. The fugacity ratio is estimated using equation 7,

$$F = \exp\left[-0.023\left(T_M - 298\right)\right] \tag{7}$$

where T_M is the melting point of the solid compound. The subcooled vapor pressure or water solubility is calculated by dividing the solid phase value by the fugacity ratio. The Henry's law values in Table 6.1 are calculated as the ratio of the subcooled vapor pressure to subcooled water solubility (VP/S). There is much uncertainty in physical and chemical property measurement data for pesticides and, generally, several different experimental values can be found in the literature for properties, such as water solubility and vapor pressure.

TABLE 6.1. National use rank in 1988, water solubility, vapor pressure, and Henry's law values for selected pesticides between 20 and 25°C

[Inorganic pesticides are not included in this table. mole/m^3, mole per cubic meter; Pa, pascal]

1988 Use rank	Compound	Subcooled liquid		Henry's law	Reference
		Water solubility (mole/m^3)	Vapor pressure (Pa)		
2	Atrazine	4.48E+00	1.29E-03	2.87E-04	Suntio and others, 1988
3	Alachlor	6.88E+01	4.14E-03	6.02E-03	Suntio and others, 1988
4	Metolachlor	1.87E+00	1.70E-03	9.10E-04	Worthing and Walker, 1987
5	EPTC	1.95E+00	2.00E+00	1.027E+00	Suntio and others, 1988
6	2,4-D (acid)	2.35E+01	1.30E+01	5.53E-01	Suntio and others, 1988
7	Trifluralin	2.44E-03	9.84E-03	4.03E+00	Suntio and others, 1988
8	Cyanazine	1.85E+01	5.21E-06	2.82E-07	Worthing and Walker, 1987
9	Butylate	1.84E-02	1.00E-01	5.44E-01	Suntio and others, 1988
10	Chlorpyrifos	1.25E-03	2.19E-03	1.75E+00	Suntio and others, 1988
11	Pendimethalin	2.18E-03	8.16E-03	3.75E+00	Worthing and Walker, 1987
12	Chlorothalonil	4.04E-01	2.32E+02	5.76E+02	Worthing and Walker, 1987
13	Glyphosate	3.97E+03	5.60E-02	1.41E-05	Montgomery, 1993
14	Dicamba	2.01E+02	2.38E-02	1.18E-04	Suntio and others, 1988
18	Methyl parathion	1.27E-01	2.67E-03	2.11E-02	Suntio and others, 1988
19	Carbaryl	2.35E+00	2.95E-03	1.26E-03	Suntio and others, 1988
20	Propanil	6.50E+00	2.36E-02	3.64E-03	Suntio and others, 1988
21	Terbufos	1.56E-02	3.51E-02	2.25E+00	Montgomery, 1993
22	Carbofuran	5.33E+01	2.72E-02	5.11E-04	Suntio and others, 1988
24	Metribuzin	4.97E+01	5.89E-04	1.18E-05	Montgomery, 1993
25	Phorate	1.54E-01	1.00E-01	6.51E-01	Suntio and others, 1988
26	Molinate	4.70E+00	7.46E-01	1.59E-01	Worthing and Walker, 1987
27	MCPA	3.53E+01	1.72E-03	4.86E-05	Worthing and Walker, 1987
28	Fonofos	5.28E-02	2.80E-02	5.30E-01	Worthing and Walker, 1987
29	Propazine	2.83E+00	2.94E-04	1.04E-04	Suntio and others, 1988
30	Propachlor	9.37E+00	9.92E-02	1.06E-02	Suntio and others, 1988
31	Simazine	2.52E+00	8.65E-04	3.43E-04	Suntio and others, 1988
32	Propargite	1.43E-03	4.00E-01	2.80E+02	Wauchope and others, 1992
33	Captan	5.61E-02	3.38E-02	6.01E-01	Suntio and others, 1988
35	Aldicarb	1.73E+02	5.48E-02	3.17E-04	Suntio and others, 1988
36	Ethafluralin	1.21E-03	2.22E-04	1.83E-01	Worthing and Walker, 1987
37	Triallate	1.09E-02	1.11E-02	1.02E+00	Suntio and others, 1988
38	Malathion	2.64E-01	6.01E-04	2.28E-03	Suntio and others, 1988
39	Disulfoton	9.11E-02	2.00E-02	2.20E-01	Suntio and others, 1988
41	Chloramben	1.92E+02	5.27E+01	2.74E-01	Worthing and Walker, 1987
43	Acephate	1.43E+04	9.09E-04	6.37E-08	Worthing and Walker, 1987
44	Dimethoate	1.62E+02	1.86E-02	1.15E-04	Suntio and others, 1988
45	Methomyl	2.11E-02	1.37E-02	6.49E-05	Suntio and others, 1988
46	Picloram	1.53E+02	5.14E-03	3.37E-05	Suntio and others, 1988
47	Parathion	3.33E-02	3.88E-04	1.17E-02	Suntio and others, 1988
50	Linuron	1.26E+00	6.77E-03	5.37E-03	Suntio and others, 1988
51	Azinphos-methyl	2.89E-01	9.15E-05	3.17E-04	Suntio and others, 1988
52	Fluometuron	1.10E+01	1.61E-03	1.46E-04	Worthing and Walker, 1987
53	Dacthal	3.06E-02	6.78E-03	2.21E-01	Wauchope and others, 1992
54	Endosulfan	1.47E-03	4.37E-03	2.98E+00	Suntio and others, 1988
55	Diuron	3.70E+00	4.31E-03	1.17E-03	Suntio and others, 1988

TABLE 6.1. National use rank in 1988, water solubility, vapor pressure, and Henry's law values for selected pesticides between 20 and 25°C--*Continued*

1988 Use rank	Compound	Subcooled liquid Water solubility (mole/m^3)	Vapor pressure (Pa)	Henry's law	Reference
57	Prometryn	1.73E+00	8.69E-04	5.03E-04	Suntio and others, 1988
58	Norflurazon	3.04E+00	9.24E-05	3.04E-05	Worthing and Walker, 1987
61	Diazinon	1.25E-01	8.00E-03	6.41E-02	Suntio and others, 1988
63	Ethoprop	2.89E+00	4.65E-02	1.61E-02	Montgomery, 1993
64	Acifluorfen	6.02E+00	2.41E-03	4.01E-04	Worthing and Walker, 1987
65	Diclofop-methyl	1.24E-02	4.80E-05	3.87E-03	Worthing and Walker, 1987
68	Thiobencarb	6.59E-02	1.78E-03	2.70E-02	Wauchope and others, 1992
70	Ethion	4.68E-03	1.50E-04	3.20E-02	Suntio and others, 1988
71	Benfluralin (Benefin)	7.61E-03	1.02E-02	1.34E+00	Suntio and others, 1988
74	Bromacil	5.53E+01	1.08E-01	1.95E-03	Suntio and others, 1988
75	Methamidophos	1.11E+04	1.67E-01	1.51E-05	Wauchope and others, 1992
76	Permethrin	3.76E-05	5.89E-06	1.57E-01	Wauchope and others, 1992
78	Terbutryn	6.37E-01	8.00E-04	1.26E-03	Suntio and others, 1988
80	Asulam	3.28E+04	2.01E-05	6.14E-10	Montgomery, 1993
84	Phosmet	1.85E-01	1.76E-04	9.52E-04	Suntio and others, 1988
87	Vernolate	4.43E-01	9.00E-01	2.03E+00	Suntio and others, 1988
88	Fenamiphos	4.01E+00	2.31E-04	5.76E-05	Worthing and Walker, 1987
89	Fluazifop-P-butyl	1.65E-03	3.41E-04	2.07E-01	Worthing and Walker, 1987
90	Oxamyl	6.55E-02	1.72E-01	2.63E-04	Suntio and others, 1988
91	Napropamide	8.49E-01	1.67E-03	1.97E-03	Worthing and Walker, 1987
94	Pebulate	2.95E-01	3.50E+00	1.19E+01	Suntio and others, 1988
97	Bensulide	4.43E-05	1.77E-02	3.99E+02	Wauchope and others, 1992
98	Profluralin	2.74E-04	1.07-02	3.91E+01	Suntio and others, 1988
99	Tebuthiuron	2.60E+02	6.42E-03	2.47E-05	Worthing and Walker, 1987
100	Oxyfluorfen	8.63E-04	8.34E-05	9.66E-02	Wauchope and others, 1992
103	Diethatyl ethyl	5.92E-01	7.50E-04	1.27E-03	Worthing and Walker, 1987
106	Dalapon	3.51E+03	1.60E+01	4.56E-03	Howard and others, 1991
108	Dinoseb	2.76E-01	1.41E+01	5.11E+01	Suntio and others, 1988
109	Methidathion	1.15E+00	2.60E-04	2.25E-04	Worthing and Walker, 1987
110	Terbacil	8.92E+01	1.61E-03	1.81E-05	Suntio and others, 1988
112	Hexazinon(e)	1.06E+03	2.19E-04	2.06E-07	Worthing and Walker, 1987
118	Methazole	5.54E-02	1.28E-03	2.32E-02	Worthing and Walker, 1987
119	Chlorimuron-ethyl	1.17E+02	2.16E-08	1.84E-10	Wauchope and others, 1992
120	Esfenvalerate	1.06E-05	3.26E-06	3.09E-01	Wauchope and others, 1992
125	Chlorpropham	6.75E-01	1.44E-03	2.14E-03	Suntio and others, 1988
126	Pronamide	1.18E+00	2.27E-01	1.93E-01	Worthing and Walker, 1987
130	Tridiphane	8.46E-03	4.37E-02	5.16E+00	Worthing and Walker, 1987
134	Cypermethrin	3.44E-05	6.69E-07	1.94E-02	Wauchope and others, 1992
135	Ametryn	3.24E+00	3.97E-04	1.23E-04	Suntio and others, 1988
136	Phenmedipham	2.39E-01	1.98E-08	8.31E-08	Worthing and Walker, 1987
142	Desmedipham	2.07E-01	3.56E-06	1.72E-05	Wauchope and others, 1992
146	Diallate	5.81E-02	1.46E-02	2.51E-01	Suntio and others, 1988
152	Lactofen	3.43E-04	1.69E-06	4.93E-03	Wauchope and others, 1992
157	Chlorsulfuron	6.31E+02	1.98E-02	3.13E-05	Wauchope and others, 1992
159	Fenvalerate	4.76E-06	1.47E-06	3.08E-01	Wauchope and others, 1992
160	Triclopyr	2.97E+01	2.91E-03	9.79E-05	Worthing and Walker, 1987

TABLE 6.1. National use rank in 1988, water solubility, vapor pressure, and Henry's law values for selected pesticides between 20 and 25°C—*Continued*

1988 Use rank	Compound	Subcooled liquid		Henry's law	Reference
		Water solubility (mole/m^3)	Vapor pressure (Pa)		
161	HCH, γ- (Lindane)	1.65E-01	2.22E-02	1.34E-01	Suntio and others, 1988
162	Chloroxuron	2.52E-01	4.39E-06	1.74E-05	Worthing and Walker, 1987
163	Dichlobenil	1.63E+00	1.09E+00	6.69E-01	Suntio and others, 1988
166	Barban	1.36E-04	1.60E-04	1.17E+00	Worthing and Walker, 1987
170	Diflubenzuron	3.68E-02	1.48E-03	4.04E-02	Worthing and Walker, 1987
171	Metsulfuron-methyl	6.16E+02	8.25E-09	1.34E-11	Wauchope and others, 1992
176	Fenoxaprop-ethyl	8.69E-03	1.68E-05	1.93E-03	Wauchope and others, 1992
177	Clopyralid	8.60E+02	2.94E-02	3.41E-05	Worthing and Walker, 1987
180	Carboxin	3.87E+00	1.12E-04	2.90E-05	Wauchope and others, 1992
	Aldrin	3.39E-04	3.09E-02	9.12E+01	Suntio and others, 1988
	Chlordane	7.51E-04	6.77E-03	9.02E+00	Suntio and others, 1988
	DDD, *p,p'*-	1.16E-03	7.40E-04	6.40E-01	Suntio and others, 1988
	DDE, *p,p'*-	5.48E-04	4.36E-03	7.95E+00	Suntio and others, 1988
	DDT, *p,p'*-	5.81E-05	1.37E-04	2.36E-00	Suntio and others, 1988
	Dieldrin	1.46E-02	1.63E-02	1.12E+00	Suntio and others, 1988
	Endrin	4.16E-02	1.38E-03	3.31E-02	Suntio and others, 1988
	HCH, α-	7.41E-02	6.47E-02	8.72E-01	Suntio and others, 1988
	HCH, β-	2.39E-01	2.72E-02	7.27E-02	Suntio and others, 1988
	HCH, δ-	3.74E-01	3.09E-02	8.25E-02	Suntio and others, 1988
	Heptachlor	1.36E-03	1.52E-01	1.12E+02	Suntio and others, 1988
	Heptachlor epoxide	1.52E-02	9.97E-01	6.55E+01	Montgomery, 1993
	Pentachlorophenol (PCP)	2.05E+00	9.10E-02	4.44E-02	Suntio and others, 1988
	Terbuthylazine	1.25E+00	5.06E-03	4.05E-03	Worthing and Walker, 1987
	Toxaphene	5.39E-03	2.23E-03	4.14E-01	Suntio and others, 1988

Transformation reactions are important to airborne pesticides because they are part of the removal process along with wet and dry deposition, and they may also result in products that are more toxic or more persistent, or both, than the parent compound. In the atmosphere, pesticides can undergo transformation reactions induced directly by sunlight (direct photolysis), reactive oxidant species created by sunlight (indirect photolysis), or condensed water (hydrolysis). Photochemical reactions are probably the most important reaction type for airborne pesticides because of the extended exposure to sunlight. The kinetics of specific reactions for individual pesticides are strongly dependent on the environmental phases in which the pesticide is present in the atmosphere. As an extreme example, a pesticide that rapidly undergoes acid-catalyzed hydrolysis will be quickly removed from the atmosphere if it is distributed to a large extent into a condensed water phase in clouds or raindrops (low Henry's law constant). If a pesticide is predominately present in the vapor phase (high Henry's low constant, low particle to gas partition coefficient, or lack of condensed water droplets in the atmosphere), then the kinetics of removal via hydrolytic reactions will be slow and a relatively unimportant factor in the atmospheric concentration. The sorption of pesticides to particulate matter may change the characteristic vapor-phase absorbance spectra. This could enhance or reduce the reactivity of the compound (Kempny and others, 1981).

Some pesticides, such as the organophosphorus insecticides parathion, methyl parathion, and fonophos can react photochemically to form the corresponding oxygen analogs in a few minutes to several hours (Woodrow and others, 1977, 1978; Klisenko and Pis'mennaya, 1979). The organophosphorus oxygen analog is more toxic than the parent thion, but in the case case of parathion, the oxygen analog can be further transformed to the phenol and phosphates (Woodrow and others, 1983). Folex, an organophosphorus cotton defoliant, was essentially completely oxidized within seconds after application, whereas the photoproduct DEF remained essentially stable (Woodrow and others, 1983). In laboratory experiments, DEF was shown to break down slowly in the presence of ozone (Moilanen and others, 1977), but these products were not detected in subsequent field measurements.

Dieldrin, an organochlorine insecticide, is photochemically transformed into the more toxic photodieldrin. Turner and others (1977) found that both dieldrin and photodieldrin volatilized from a grass surface, but that photodieldrin volatilized at a much slower rate. They measured the air concentration of both compounds at five heights above the field surface and calculated the ratio of photodieldrin to dieldrin air concentrations. These ratios showed that the majority of photodieldrin formed at the surface and volatilized, rather than being formed in the atmosphere, at least within the first 2 m above the surface. Laboratory experiments have shown that aldrin can be photochemically oxidized to dieldrin, which can be further photoisomerized to photodieldrin (Crosby and Moilanen, 1974). DDT has also been shown to react photochemically to produce DDE (Crosby and Moilanen, 1977). Much of the past and current research on pesticides in the atmosphere has focused on parent compounds and has virtually ignored their transformation products. Pesticide transformation products may be a significant key in more fully understanding the environmental fate of airborne pesticides.

6.3 RELATIVE IMPORTANCE OF WET AND DRY DEPOSITION

Wet deposition includes deposition to the earth's surface of airborne pesticide vapors that have partitioned into falling droplets or snowflakes as well as particle-bound pesticides that the droplets condensed around or intercepted on their way down. Wet deposition occurs during rain, snow, and fog and, possibly, with dew formation. Dry deposition includes deposition to the earth's surface of airborne pesticide vapors and particle-bound pesticides. Dry deposition is a continuous, but slow process. The relative importance of wet versus dry deposition depends upon the frequency of occurrence and the intensity of precipitation and fog events as well as the concentration of pesticides in air, the particle size distribution and concentration, and the efficiency of the removal process.

The contribution of total dry deposition (particle and gas) to the total deposition burden is largely unknown. Direct measurement of dry deposition rates of air pollutants is difficult, and the results have a high degree of uncertainty associated with them (Sehmel, 1980; Droppo, 1985; Businger, 1986; Bidleman, 1988; Sirois and Barrie, 1988). Much work has been done on measuring the dry deposition of inorganic constituents associated with particulate matter (Högström, 1979; Dasch, 1985; Lindberg and Lovett, 1985; Sievering, 1986; Likens and others, 1990; Sickles and others, 1990); however, little has been done in this area with respect to pesticides associated with particle deposition (Bidleman, 1988) and even less on pesticide vapor deposition.

Calculations by Bidleman (1988) show that the amount of organic chemical with $P_L° \approx 10^{-4}$ Pa associated with particulate matter ranges between 5 and 65 percent for ϕ values (equation 1) from clean air to urban air environments. The concentration and composition of the atmospheric particles, however, determine the actual distribution and atmospheric residence times of most pesticides (Eisenreich and others, 1980; Pankow, 1987; Bidleman, 1988). Vapor-particle partitioning in the atmosphere is dependent upon the pesticide vapor pressure, and the size, surface area, and organic content of the suspended particles (Junge, 1977; Pankow, 1987). In general, those compounds with $P_L° \geq 0.05$ Pa are expected to exist almost entirely in the vapor phase, whereas those compounds with $P_L° \leq \approx 10^{-6}$ Pa should exist almost entirely sorbed to particulate matter. In fact, most pesticides lie somewhere between this range of $P_L°$ values at 20°C (see Table 6.1). Eisenreich and others (1981) estimated that total dry deposition can be 1.5 to 5 times that of wet deposition for organochlorine compounds to the Great Lakes. These values, however, are dependent on the uncertainty of estimating actual dry deposition rates.

Scavenging of airborne pesticide vapors by water droplets is favored by those compounds with low Henry's law values (H), which tend to partition into the water droplet. In addition, these compounds also tend to partition onto airborne particles, which can then be removed by wet and dry deposition. H values alone do not always satisfactorily describe this air-water equilibrium. Differences have been attributed, in part, to the strong temperature dependence of the H values, which can change by a factor of two for each 10°C temperature change (Ligocki and others, 1985a; Larsson and Okla, 1989). It also may be due to particle scavenging by the falling raindrops.

Actual field measurements have shown that measured pesticide vapor washout values (W_g) are often higher than the calculated ones and can differ from one to several orders of magnitude. This is often due to inaccuracies in H values especially at environmentally relevant temperatures (Pankow and others, 1984, Ligocki and others, 1985a). To further complicate matters, surface active organic material can form surface films on rain and fog droplets and snowflakes. These surface films can retard the evaporation of the water droplet and the solubilized pesticide. They also can reduce the diffusion rate of water soluble compounds from the ambient air into the droplet (Gill and others, 1983; Graedel and others, 1983). Organic films around the droplet ultimately affect the equilibrium process (Giddings and Baker, 1977; Capel and others, 1990, 1991), which can enrich or diminish concentrations from the expected equilibrium values.

The importance of wet versus dry deposition of pesticides depends upon the frequency, intensity, and duration of precipitation events. Precipitation cleans the atmosphere of most pollutants. Most oxidative reaction products are more polar than the parent compound, which suggests that they will also be more water soluble and more readily removed by wet depositional processes. This can happen in areas with frequent precipitation like the Pacific Northwest, Midwest, and eastern United States. In the arid Southwest, this may not be the case. Photoreactions may be more important in these dry areas since the time between precipitation events is generally long and airborne concentrations of ozone and particulate matter can become high, which can enhance reaction rates.

6.4 SAMPLING METHOD EFFECTS ON APPARENT PHASE DISTRIBUTIONS

The goal of sampling environmental matrices is to collect a representative sample. Often, the act of collecting the sample can change the distribution of the compound between different matrices if certain precautions are not followed. In studies designed to determine the distribution of pesticides, or any other semivolatile organic compound between the vapor and adsorbed particle phases, special care must be taken when collecting the sample so as not to affect the natural distribution of the pesticides.

Pesticides exist in the atmosphere as vapors and sorbed to particles. These two phases are often sampled together and differentiated to determine the distribution of the pesticide and the dominant removal process. A number of factors influence the distribution of a particular pesticide between vapor and particles, such as the length of the sampling period, the vapor phase concentration, the concentration of particulate matter, and the ambient temperature and humidity (Bidleman and others, 1986; Bidleman and Foreman, 1987; Ligocki and Pankow, 1989; Pankow and Bidleman, 1991). During field-scale studies of worker exposure, drift, and post-application volatilization of a test compound, the distinction between vapor and particle usually is not made because the major part of the air concentration is usually associated with the vapor phase. The sampling periods are usually short, on the order of 0.5 to 4 hours, and the amount of particulate matter collected during this time is usually negligible. Those studies monitoring the distribution of trace level of pesticides in the ambient air at sites removed from major sources usually require longer sampling times, on the order of 24 or more hours per period. These long sampling times are required to collect enough material for analysis. During long sampling periods, several factors can affect the sample. Free vapor can sorb to the trapped particles as well as to the filter (adsorption gains), resulting in an artificially high particle phase concentration (Foreman and Bidleman, 1987; McDow and Huntzicker, 1990; Zhang and McMurry, 1991; Cotham and Bidleman, 1992; Hart and others, 1992). Sorbed material can also desorb from the trapped particulate matter (blow-off losses) and be trapped on the vapor sorbent, resulting in an inflated vapor phase concentration. The latter is the most common sampling artifact reported (Bidleman and Olney, 1974; Harvey and Steinhauer, 1974; Grosjean, 1983; Spitzer and Dannecker, 1983; Van Vaeck and others, 1984), and much work has been done on sampler design to reduce this (Appel and others, 1983; Lane and others, 1988; Coutant and others, 1988, 1989; Cotham and Bidleman, 1992; Hart and others, 1992; Krieger and Hites, 1992; Turpin and others, 1993).

Long sampling periods (24 hours or more) have the additional complication of changing ambient temperatures and humidity throughout the sampling period. Temperature can enhance or diminish adsorption gains and blow-off losses because of its effect on vapor pressure (Bidleman and Foreman, 1987). As the water vapor content of the atmosphere increases, the percentage of pesticide (or other semivolatile organic compounds) sorbed to the particulate material decreases (Chiou and Shoup, 1985; Thibodeaux and others, 1991; Goss, 1993; Pankow and others, 1993). This is similar to the observed behavior of pesticides in soil (Spencer and others, 1982).

Oxygen analog formation from organophosphorus pesticides can occur on the adsorbent as air passes through it during high-volume air sampling, but this does not seem to be a severe problem if the collection media is shielded from direct sunlight (Woodrow and others, 1977; Seiber and others, 1989).

Rain sampling is straightforward for ground-based sampling, but the sampler design and how the sample is collected can play a critical role in how the results are interpreted. Most of these samplers usually have a large collecting area that drains into a bucket or through an extraction cartridge of some sort. The collecting efficiencies of several commercially available models have been reviewed by Franz and others (1991). Some remote sampling devices sample on an event basis. They are covered during the dry periods and have a moisture sensor that opens the cover during precipitation events similar to those samplers used by Harder and others (1980), Pankow and others (1984), Zabik and Seiber (1992), and Goolsby and others (1994). Another type of sampler is continuously open to the environment. The precipitation collected with these samplers, if they are not cleaned prior to the precipitation event, will reflect the dry deposition as well as the rain. These samples can be used to estimate bulk deposition for the exposed period; similar samplers were used by Harder and others (1980), Glotfelty and others (1990b), and

Nations and Hallberg (1992). These types of continuous deposition samplers cannot be used to estimate rain-only deposition unless they are cleaned just prior to the onset of a rain event.

Fog sampling must efficiently collect a representative distribution across the droplet diameter spectrum (1 to 100 μm) while avoiding collecting submicron aerosols (Jacob and others, 1983, 1985) and evaporating any part of the collected water. Discrimination against the smaller droplet diameters may result in lower than actual concentrations because of dilution from the larger diameter droplets. Likewise a discrimination against the larger diameter droplets may result in higher than actual concentrations because of the lower water content of the smaller diameter droplets. The object of fog samplers is to preserve the size and chemical composition of the fog droplets throughout all stages of the collection (Jacob and others, 1985) and to minimize fog droplet evaporation during the sampling process. During sampling, the larger droplets can impact the collecting surface and disintegrate into smaller ones having greater surface area and a greater evaporation potential. The smaller droplets could also pass through the collecting surface uncollected.

Water vapor normally condenses around particulate matter in the formation of rain and fog droplets. Rain droplets can also intercept and incorporate airborne particles as well as vapors as they fall to earth (see Figure 4.4). Partitioning between the vapor, liquid, and particle phases occurs as the rain or fog droplets fall to the ground. This partitioning can continue in the bulk water sample after it is collected until an equilibrium is reached. In order to preserve the actual water-particle concentration distribution as much as possible, the rain and fog water must be filtered through a highly efficient filter as it is collected or shortly afterwards. Analysis of unfiltered rain and fog water will result in total pesticide deposition.

Sampling dry deposition, both particulate material and vapors, by passive plate collectors coated with a sticky material or water treats the deposition as "ideal." That is, the sampler does not account for rebound and reentrainment of the deposited material, which can in some cases reduce the net deposition to zero (Paw U, 1992). In these cases, the results can be erroneously high. Bidleman (1988), however, stated that pesticide vapors should not sorb to the coated (sticky) collectors. These conflicting views of sampling dry deposition are indicative of the state of knowledge on sampling dry deposition, and much more work needs to be done to settle the issue. In all sampling methods for air, particulate matter, fog, or precipitation, any partitioning or sorption of the pesticide to the sampler itself must be minimized. Stainless steel, aluminium, or Teflon are the preferred construction material.

CHAPTER 7

Analysis of Key Topics: Environmental Significance

The presence of pesticides in the atmosphere can have environmental significance. It has been shown that airborne pesticides can be transported from their application site and deposited in areas many kilometers away where their use was not intended. Atmospheric deposition of pesticides can have an effect on water quality, fish and other aquatic organisms within the affected body of water, and on humans that consume affected fish.

7.1 CONTRIBUTION TO SURFACE- AND GROUND-WATER

The potential contribution of pesticides from the atmosphere to a surface-water body depends on pesticide levels in atmospheric deposition and on how much of the water budget is derived from surface runoff and direct precipitation. Therefore, the relative importance of atmospheric inputs to surface waters compared to other nonpoint sources is, generally, proportional to the surface area of the body of water compared to its terrestrial drainage area. For example, a lake with a large surface area with respect to its drainage area, such as Lake Superior, usually receives much of its total inflow of water from direct precipitation and is vulnerable to atmospheric contaminants. In contrast, a stream draining a basin with low relief and permeable soils usually receives only minor contributions from direct precipitation of surface runoff, although such contributions may be great during intense storm events. A small stream draining an urban area or other areas with high proportions of impervious surface in its drainage basin may yield streamflow during storm events that is largely comprised of precipitation and direct surface runoff. Few systems have been studied, however.

Most studies of atmospheric deposition of pesticides to surface water have been for selected organochlorine pesticides in the Great Lakes. Strachan and Eisenreich (1990) estimated that atmospheric deposition is the greatest source of PCB and DDT input into Lakes Superior, Michigan, and Huron. Murphy (1984) used precipitation concentration data from Strachan and Huneault (1979) to estimate the loadings of eight organochlorine pesticides into four of the Great Lakes for 1975-76. The depositional amounts ranged from 112 kg/yr for HCB to nearly 1,800 kg/yr for α-HCH, roughly the same as reported by Eisenreich and others (1981). Strachan (1985) reported that the precipitation inputs at two locations at opposite ends of Lake Superior contained a variety of organochlorine pesticides. The calculated average yearly loadings ranged from 3.7 kg/yr for HCB to 860 kg/yr for α-HCH (Table 7.1). The loading estimates noted in Table 7.1 show greater input from dryfall, but this is because rain events occur less frequently. Voldner and

TABLE 7.1. Estimates of rainfall loadings of organics to Lake Superior in 1983

[km^2, square kilometer; mm, millimeter; ng/L, nanogram per liter; kg/yr, kilogram per year; --, no data. Previous estimate data for rain from Science Advisory Board, 1980, Table 27, and for dryfall from Eisenreich and others, 1980, Table 7]

Compound	Volume weighted rain concentration[1] (ng/L)	Loadings from rain and snow[2] (kg/yr)	Previous estimates (kg/yr) Rain	Previous estimates (kg/yr) Dryfall
α-BHC (i.e., α-HCH)	17.4	860	1,000	2,300
Lindane (γ-HCH)	5.9	290	330	15,600
Heptachlor epoxide	0.35	17.0	--	
Dieldrin	0.56	28.0	130	410
Endrin	0.085[3]	4.2	--	
p,p'-DDE	0.12[3]	5.9		
p,p'-DDT	0.11	5.4	} 330	} 250
p,p'-DDD	0.11[3]	5.4 } 17.0		
Methoxychlor	2.4	120	520	7,,800
PCBs	6.0	300	2,000	7,800
HCB	0.075	3.7	130	1,600

[1]One-half the detection limit was used when no compound was detected.
[2]Rain, 580 mm; snowmelt, 225 mm; surface area of lake, 82,100 km^2.
[3]Less than one-half of the samples contained this compound.

Reprinted with permission from Environmental Toxicology and Chemistry, Volume 4(5), W.M.J. Strachan, Organic Substances in the Rainfall of Lake Superior: 1983, Copyright 1985 SETAC.

Schroeder (1989) estimated that 70-80 percent of the toxaphene loading to the Great Lakes was derived from long-range transport and wet deposition. This included inputs from secondary sources such as revolatilization, resuspension, and runoff resulting from atmospheric deposition to the basins surrounding the Lakes.

Very little research has been done on the depositional inputs of pesticides into surface waters outside the Great Lakes area or for pesticides other than organochlorine compounds. Capel (1991) estimated the yearly wet depositional fluxes of alachlor, atrazine, and cyanazine in Minnesota to be on the order of 40, 20, and 20 metric tons, respectively. These values represent approximately 1 percent of the total applied for each compound in Minnesota. What is not known is the unintended herbicidal effects these chronic depositional levels have on the flora of terrestrial and aquatic areas, or even how accurate these depositional estimates are. Wu (1981) estimated that the atrazine inputs into a small watershed-estuary system of the Rhode River on Chesapeake Bay, Maryland, to be 1,016 and 97 mg/ha in 1977 and 1978, respectively. The reasons for the 10-fold difference in calculated loadings between the two years may have been due to long-range transport of polluted air masses into the area. Glotfelty and others (1990c) estimated that approximately 3 percent of the atrazine concentration and 20 percent of the alachlor concentration found in the Wye River, on Chesapeake Bay, was attributable to precipitational inputs. They also estimated that the average summer wet deposition inputs into Chesapeake Bay for atrazine, simazine, alachlor, metolachlor, and toxaphene were 0.91, 0.13, 5.3, 2.5, and 0.82 metric tons, respectively, between 1981 and 1984. However, these estimates were made with the assumptions that the pesticide air concentrations were uniform over the entire 11.9×10^4 m^2 area of the Bay, and that the rainfall was also uniform across the Bay. Direct vapor-water partitioning was not accounted for, and these values are, most likely, conservatively low.

There are several reasons why the importance of atmospheric deposition of pesticides into surface waters is largely unknown. Eisenreich and others (1981) listed them more than a decade ago and they still hold true today. They are:

(1) Inadequate database on atmospheric concentrations of pesticides.
(2) Inadequate knowledge of pesticide distribution between vapor and particle phases in the atmosphere.
(3) Lack of understanding of the dry deposition process.
(4) Lack of appreciation for the episodic nature of atmospheric deposition.
(5) Inadequate understanding of the temporal and spatial variations in atmospheric concentration and deposition of pesticides, and as Bidleman (1988) noted,
(6) Incomplete or questionable physical property data.

The potential contribution of pesticides from the atmosphere to ground water depends on the pesticide levels in atmospheric deposition and on the portion of ground-water recharge that is derived from precipitation. The actual contribution of airborne pesticides to ground water is strongly affected by the degree of filtering and sorption of pesticides that occurs as infiltrating precipitation passes through the soil and underlying unsaturated zone to the water table. The extent of sorption depends on the degree of contact with the soil and on the chemical properties of both the pesticide and the soil.   The greatest contribution of pesticides from the atmosphere is likely to occur when precipitation is the major source of recharge and the unsaturated zone is highly permeable, particularly if there are macropores, cracks, or fissures in the soil (Shaffer and others, 1979; Thomas and Phillips, 1979; Simson and Cunningham, 1982).

Studies done in the United States that investigated ground-water contamination by pesticides in precipitation recharge are few, if any. Schrimpff (1984) investigated the precipitation input of α- and γ-HCH, and several PAHs into two Bavarian watershed ground-water systems (the ancient earthblock and the scarplands) and found that only one percent of the α- and γ-HCH percolated into the shallow ground water. He concluded that the soil above the water table was effective in filtering the recharge water. Simmleit and Herrmann (1987a,b) also investigated the contamination of Bavarian ground water by α- and γ-HCH and several PAHs from snowmelt in a very porous karst ground-water system. They found that from an average bulk precipitation γ-HCH concentration of about 40.0 ng/L, the concentration of trickling water at depths of 2 m, 7 m, and 15 to 20 m were 0.2, 0.1 ng/L, and none detected, respectively. These studies show that the soil in these areas is a good filter for γ-HCH, an organochlorine insecticide. Contamination of ground water by pesticides with greater solubility in water does occur, but how much of this contamination can be attributed to atmospheric deposition is not known.

7.2 HUMAN HEALTH AND AQUATIC LIFE

The most clearly documented effects of pesticides in the atmosphere on human health and aquatic life are related to long-lived, environmentally stable organochlorine insecticides that concentrate in organisms through biomagnification (food chain accumulation), bioconcentration (partitioning), or both. Through these processes, organochlorine insecticides, even at the low levels frequently found in air, rain, and fog, have been found to concentrate to significant levels in fish, mammals and humans.

The U.S. Fish and Wildlife Service periodically monitors the concentrations of organochlorine compounds in freshwater fish from a network of over 100 stations nationwide. Their analyses cannot determine the source of the contamination or determine how much is derived from atmospheric deposition, but Schmitt and others (1983) found α-HCH residues in fish throughout the country and speculated that the major source of this contamination resulted from atmospheric transport and deposition. In particular, as discussed in the previous section on the contribution of atmospheric deposition of pesticides to surface-water sources, several of the Great Lakes, and especially Lake Superior, derive most of their organochlorine contamination from atmospheric deposition, with toxaphene being the most notable example. Between 1977 and 1979 toxaphene concentrations in whole fish, mostly lake trout (*Salvelinus namaycush*) and bloater (*Coregonus hoyi*), frequently exceeded the Food and Drug Administration (FDA) action level of 5.0 mg/kg wet weight, which was set for the edible portions of fish (Rice and Evans, 1984). Since then, however, toxaphene and most other organochlorine concentrations in fish have been decreasing (Schmitt and others, 1990) in correspondence with reduced North American use, but there still exist many other sources for these pesticides worldwide.

Determining the significance to human health and aquatic life of non-organochlorine pesticides in air, rain, snow, and fog is not straightforward because there are no existing national standards or guidelines for these matrices and other pesticides do not persist to the same degree as organochlorine insecticides. Nevertheless, a general perspective on the potential significance is aided by comparing rain water concentrations to standards and guidelines for water. The USEPA has set standards and guidelines for contaminant levels that may occur in public water systems that can adversely affect human health, which include the regulatory MCL (Maximum Concentration Level) and the 1-day and long-term exposure health advisories for children (U.S. Environmental Protection Agency, 1994a). In addition to human health concerns, there are USEPA and NAS (National Academy of Sciences) water-quality criteria for protection of aquatic organisms (U.S. Environmental Protection Agency, 1994a; National Academy of Sciences/ National Academy of Engineering, 1973), which are often more sensitive to low-level pesticide exposures than are humans.

Table 7.2 lists these values, where available, for those pesticides that have been analyzed for in the atmosphere at 10 or more sites in the United States, along with the range of concentrations and matrix in which they were detected. Only 25 percent of the pesticides analyzed for in the various atmospheric matrices have associated MCL values, about 57 percent have a child long- or short-term health advisory value, 44 percent have TWA (time-weighted average) values, and about 32 percent have aquatic-life criteria values. Only chlordane, endrin, and heptachlor have values for each of these criteria.

In most cases the measured pesticide concentrations in rain are one or more orders of magnitude below the human-health related values for drinking water. There are several instances, though, where the concentrations in rain have exceeded the MCL values. These have occurred for alachlor, atrazine, and 2,4-D. Cyanazine, 2,4-D, and 2,4,5-T exceeded, and atrazine has been detected in several samples near the long-term exposure limit for children. In general, the very high concentrations measured in rain occurred infrequently. They occurred in or near agricultural areas where pesticides were applied and could be due to unusual circumstances resulting in abnormally high concentrations, such as a brief but small amount of rainfall during or soon after an application to a large area. A study that measured the concentrations of several pesticides in residential, office, and warehouse air during applications to lawns, trees, and shrubs (Yeary and Leonard, 1993) found that about 80 percent of the 500 samples collected were below the detectable limits of 0.001 mg/m^3. Of the pesticides that were detected, the TWA values were generally less than 10 percent of any standard (Yeary and Leonard, 1993).

TABLE 7.2. Water- and air-quality criteria for humans and aquatic organisms and the concentration range at which each pesticide was detected (if detected) in rain, air, fog, and snow

[ng/L, nanogram per liter; ng/m³, nanogram per cubic meter; USEPA, U.S. Environmental Protection Agency; MCL, maximum contaminant level; ND, not detected; OA, oxygen analog transformation of the parent compound; TWA, time-weighted average; NAS, National Academy of Sciences; nsg, no standard or guideline exists for this compound; <, less than; --, no data; †, primary drinking water regulations; ‡, drinking water health advisories; ¥, air-quality criteria limits for air contaminants (Occupational Safety and Health Administration, 1989). Water-quality criteria are from a compilation of national standards and guidelines for pesticides in water by Nowell and Resek, 1994]

Compound	Water Quality Criteria, humans: USEPA drinking water values			Air	Water Quality Criteria, aquatic organisms: freshwater (ng/L)			Observed concentrations							
	MCL† (ng/L)	Child (ng/L)‡		Adult TWA¥ (ng/m³)	USEPA		NAS	Rain (ng/L)		Air (ng/m³)		Fog (ng/L)		Snow (ng/L)	
		1-day	Long-term		Acute	Chronic		Low	High	Low	High	Low	High	Low	High
Alachlor	2,000	100,000	100,000	nsg	nsg	nsg	nsg	10	22,000	0.06	42.9	1,450	--	--	--
Aldrin	nsg	300	300	250,000	3,000	nsg	10	0.01	3.4	0.1	150	--	--	--	--
Ametryn	nsg	9,000,000	900,000	nsg	nsg	nsg	nsg	ND	--	--	--	--	--	20	30
Atrazine	3,000	100,000	50,000	5,000,000	nsg	nsg	nsg	3	40,000	0.008	20	270	820	--	--
Azodrin	nsg	nsg	nsg	nsg	nsg	nsg	nsg	--	--	0.4	14	--	--	--	--
Carbaryl	nsg	1,000,000	1,000,000	5,000,000	nsg	nsg	20	--	--	--	--	69	4,000	--	--
Chlordane	2,000	60,000	500	500,000	2,400	4.3	40	0.01	9.1	0.013	204	--	--	0.02	0.7
Chlorpyrifos	nsg	30,000	30,000	200,000	83	41	1	1.3	180	0.005	199	1.3	14,200	--	--
Cyanazine	nsg	100,000	20,000	nsg	nsg	nsg	nsg	20	28,000	--	--	--	--	--	--
Dacthal	nsg	80,000,000	5,000,000	nsg	nsg	nsg	nsg	--	--	0.5	2.1	--	--	--	--
DDDs	nsg	nsg	nsg	nsg	600	3,600	6	0.03	0.5	0.024	570	--	--	0.02	0.05
DDEs	nsg	nsg	nsg	nsg	1,050,000	nsg	nsg	0.02	5	0.0001	131	--	--	0.1	1.9
DDTs	nsg	nsg	nsg	1,000,000	1,100	1	2	0.01	150	0.0005	1,560	--	--	0.1	1.9
DEF/Folex	nsg	nsg	nsg	nsg	nsg	nsg	nsg	--	--	0.03	16	--	--	--	--
Diazinon	nsg	20,000	5,000	100,000	nsg	nsg	9	1.3	2,000	0.001	306.5	140	76,300	--	--
Diazinon-OA	nsg	nsg	nsg	nsg	nsg	nsg	nsg	1.3	115.8	0.0014	10.8	1.9	28,000	--	--
Dieldrin	nsg	500	500	250,000	2,500	1.9	5	0.01	30	0.0001	93	--	--	0.2	1.4
2,4-D	70,000	1,100,000	100,000	10,000,000	nsg	nsg	3,000	50	204,000	1.15	1,410	--	--	--	--
Endosulfans	nsg	nsg	nsg	100,000	220	56	3	0.1	12	0.0001	2,257	--	--	0.1	1.34
Endrin	2,000	20,000	4,500	100,000	180	2.3	2	0.04	1	0.1	59	--	--	--	--
EPTC	nsg	nsg	nsg	nsg	nsg	nsg	nsg	100	2,800	--	--	--	--	--	--
HCB	1,000	50,000	50,000	nsg	250,000	nsg	nsg	0.01	4	0.02	0.72	--	--	0.05	0.1
HCH, α-	nsg	nsg	nsg	nsg	100,000	nsg	nsg	0.4	145	0.016	10	--	--	0.43	9.8
HCH, β-	nsg	nsg	nsg	nsg	nsg	nsg	nsg	--	--	0.2	49.4	--	--	--	--
HCH, δ-	nsg	nsg	nsg	nsg	nsg	nsg	nsg	--	--	--	9.9	--	--	--	--

TABLE 7.2. Water- and air-quality criteria for humans and aquatic organisms and the concentration range at which each pesticide was detected (if detected) in rain, air, fog, and snow—Continued

Compound	Water Quality Criteria, humans: USEPA drinking water values			Air	Water Quality Criteria, aquatic organisms: freshwater (ng/L)			Observed concentrations								
	MCL† (ng/L)	Child (ng/L)‡		Adult TWA¥ (ng/m³)	USEPA		NAS	Rain (ng/L)		Air (ng/m³)		Fog (ng/L)		Snow (ng/L)		
		1-day	Long-term		Acute	Chronic		Low	High	Low	High	Low	High	Low	High	
HCH, γ-	200	1,200,000	33,000	500,000	2,000	nsg	nsg	0.3	70	0.001	107	--	--	0.1	5.3	
Heptachlor	400	10,000	1,500	500,000	520	3.8	10	0.01	0.4	0.09	19.2	--	--	0.1	19.2	
Heptachlor epoxide	200	10,000	100	nsg	520	3.8	nsg	1.8	0.03	0.002	0.1	--	--	0.03	0.4	
Kelthane	nsg	nsg	nsg	nsg	nsg	nsg	nsg	--	--	--	9.5	--	--	--	--	
Leptophos	nsg	nsg	nsg	nsg	nsg	nsg	nsg	--	--	326	1,159	--	--	--	--	
Malathion	nsg	200,000	200,000	5,000,000	nsg	100	8	10	170	0.02	270	70	2,740	--	--	
Methidathion	nsg	nsg	nsg	nsg	nsg	nsg	nsg	--	--	0.01	23.8	0.04	15,500	--	--	
Methoxychlor	40,000	6,400,000	500,000	5,000,000	nsg	30	5	0.4	38	--	--	--	--	0.1	5.8	
Methyl parathion	nsg	300,000	30,000	200,000	nsg	nsg	nsg	100	2,770	0.02	2,060	1,210	--	--	--	
Metolachlor	nsg	2,000,000	2,000,000	nsg	nsg	nsg	nsg	46	3,000	0.07	9.7	1,960	--	--	--	
Metribuzin	nsg	5,000,000	300,000	5,000,000	nsg	nsg	nsg	100	1,200	--	--	--	--	--	--	
Parathion	nsg	nsg	nsg	100,000	65	13	0.4	1.3	7,600	0.02	1,423	1500	91,400	--	--	
Parathion-OA	nsg	nsg	nsg	nsg	nsg	nsg	nsg	1.3	2,600	0.0014	40.5	11	184,000	--	--	
Pendimethalin	nsg	nsg	nsg	nsg	nsg	nsg	nsg	100	1,500	0.64	3.6	1,370	3,620	--	--	
Phorate	nsg	nsg	nsg	50,000	nsg	nsg	nsg	--	--	1.2	15	--	--	--	--	
Prometon	nsg	200,000	200,000	nsg	nsg	nsg	nsg	<50	200	--	--	--	--	--	--	
Prometryn	nsg	nsg	nsg	nsg	nsg	nsg	nsg	ND	--	--	--	--	--	--	--	
Propazine	nsg	1,000,000	500,000	nsg	nsg	nsg	nsg	<50	120	--	--	--	--	--	--	
Simazine	4,000	500,000	50,000	nsg	nsg	nsg	10,000	0.86	1,500	0.003	2	45	1,200	--	--	
Terbutryn	nsg	nsg	nsg	nsg	nsg	nsg	nsg	ND	--	--	--	--	--	--	--	
Toxaphene	3,000	500,000	nsg	500,000	730	0.2	10	0.5	497	0.01	2,520	--	--	0.085	1.7	
Trifluralin	nsg	30,000	30,000	nsg	nsg	nsg	100	80	970	0.5	63	--	--	--	--	
T, 2,4,5-	nsg	800,000	300,000	10,000,000	nsg	nsg	nsg	1,000	590,000	12	900	--	--	--	--	

To put these measured high concentrations into the proper perspective, the frequency distribution of concentrations must be known. In order to determine the concentration frequency distribution for each of the pesticides analyzed for in the United States, however, a much more complete data set is needed than is readily available from the published literature. One large-scale regional study by Goolsby and others (1994) calculated the concentration distribution in rain for one year for several herbicides used in corn and soybean production. They found that, of the 13 herbicides and selected metabolites analyzed for, 10 were detected consistently at concentrations of 100 to 200 ng/L or greater in 6,100 rain samples, but the median concentrations were below the reporting limit of 50 ng/L. The maximum atrazine, alachlor, metolachlor, and cyanazine concentrations were 10,900, 3,200, 3,000, and 2,000 ng/L, respectively, but the 99th percentile concentrations were 1,100, 970, 680, and 290 ng/L, respectively. The corresponding MCLs for atrazine and alachlor are 3,000 and 2,000 ng/L, respectively, and there are no current MCLs for metolachlor and cyanazine. These data show that only one percent of the 1,848 rain samples analyzed exceeded human health criteria for drinking water for atrazine and alachlor. Only 1 of the 13 herbicides that Goolsby and others (1994) analyzed for in rain has a water-quality criterion for aquatic organisms. This is simazine, with a value of 10,000 ng/L. Simazine had a maximum concentration of 1,500 ng/L and a 99th percentile concentration of 70 ng/L. Both of these values are well below the set limit.

Measured pesticide concentrations in fog were frequently higher than in rain, in the thousands of nanograms per liter range. Fourteen of the 48 pesticides listed in Table 7.2 were detected in fog. Only diazinon, however, was near or exceeded the human health limits for water in 5 of 24 fog events. Six pesticides, carbaryl, chlorpyrifos, diazinon, malathion, methidathion, and parathion, frequently exceeded both USEPA and NAS water-quality criteria for aquatic organisms. Movement of airborne pesticides and deposition by fog may be an important source of pesticide input to small lakes and reservoirs in or near agricultural areas in addition to being a source of contamination on nonregistered crops.

The Occupational Safety and Health Administration (OSHA) also has set limits for the airborne pesticide exposure in the workplace in TWA concentrations (Occupational Safety and Health Administration, 1989). Measured air concentrations never exceeded TWA values. These TWAs, however, are based on an 8-hour workday and a 40-hour workweek exposure in the production of pesticides and they do not take into account any chronic, low-level exposure to the pesticide.

CHAPTER 8

Summary and Conclusions

The 130 studies reviewed show that pesticides have been detected in the atmosphere within the United States and Canada since the mid-1950's. Air was the primary atmospheric matrix sampled and analyzed during the 1950's, 1960's, and 1970's, but attention shifted to analysis of precipitation and fog during the late 1970's and 1980's. A great deal of effort was expended on studying organochlorine insecticides since the mid-1970's even though many have been banned or their use greatly restricted in the United States. During the 1970's, the organophosphorus insecticides and triazine herbicides in air and rain were the focus of several studies. The 1980's brought an increase in the number of studies analyzing for these two classes of pesticides, but they were relatively few when compared to the study of the organochlorine insecticide class.

Most studies have been concentrated in a few areas of the United States, such as the Great Lakes, the northeastern states, and California. Organochlorine compounds were the general focus of studies of the Great Lakes and Canada, whereas atrazine and several other corn herbicides were the main interest in the Midwest and Northeast. A wider variety of pesticides, including organophosphorus insecticides and various herbicides, was looked for and found in California. Only two studies had sampling sites in 20 or more states, and these were done about 20 years apart. The more recent of these studies, done in 1990 to 1991, analyzed only rain for those herbicides used in corn and soybean production in the north central and northeastern states. The older study, done in 1970 to 1972, did multiresidue analyses that included a variety of organochlorine and several organophosphorus insecticides and several chlorophenoxy herbicides in air throughout the United States. The Canadian studies generally focused on the organochlorine pesticides, with the exception of several that monitored the occurrence of selected herbicides in Saskatchewan and Ontario.

Studies of pesticides in the atmosphere have encompassed a wide range of spatial scales. Local-scale studies include process and matrix distribution studies and field studies that monitor agricultural worker exposure, pesticide drift during application, and the volatilization and off-site drift of applied pesticides after application. These studies usually had very high sampling frequencies that lasted several days to several weeks and generated hundreds of samples. The analytical methods usually were specific and optimized for one or several known compounds. State, multistate, and national-scale studies include sampling locations throughout a state, large region, or the entire nation. These studies analyzed for one or more specific compounds used extensively in the study area, or did multiresidue/multiclass screening for a wide variety of compounds. These regional studies generally lasted for one or more years and generated hundreds of samples and required reliable, identical sampling methods at every location. These studies usually did not provide much detailed information on the long-range transport of pesticides as the samples were often weekly or monthly composites. They did, however, provide a coarse indication of the distribution of the types of pesticides and their occurrence in the study area.

In the local, regional, and long-range type studies, knowledge of the pesticide types and use patterns are important in designing the sampling and analytical strategy. All study types can be designed to elucidate the spatial and temporal trends of one or many different pesticides. Generally, however, there was no consistency in sampling methodologies, sampling site placement, or collection timing and duration in the studies reviewed. There was also no consistency in the selected analytes, analytical methods, or detection limits. Often, only the compounds that were detected were reported and those compounds that were analyzed for and not detected were usually not reported.

The evidence from the reviewed literature shows that most of the pesticides that have been targeted for analysis have been detected in at least one atmospheric matrix. There are many more pesticides with similar physical and chemical properties as those detected, but that have not been found in the atmosphere. This does not mean, however, that the majority of pesticides used are not present in the atmosphere. There are several reasons why a particular pesticide has not been found and these may include low use, short atmospheric residence time (considering deposition and transformation), the timing of the sampling relative to the timing of use, the predominant atmospheric phase in which it will accumulate relative to the phase being sampled, or, perhaps most important, whether or not it has been analyzed for in the atmosphere.

The pesticides detected in the reviewed studies fall into four main categories-- organochlorine insecticides, organophosphorus insecticides, triazine and acetanilide herbicides, and other herbicides. Organochlorine insecticides, because of their widespread use during the 1950's, 1960's, and 1970's and their resistance to environmental transformations, were detected in the atmosphere of every state in which they were analyzed for. Organophosphorus insecticides also have been heavily used for decades and are still in high use, although the amounts used are decreasing. As a class, they are not as environmentally persistent as the organochlorine compounds, but they are still widely used and have been detected in most states. They have not been included in the analyte list of most studies, however. Triazine herbicides have been in use since the 1960's, but studies analyzing for these compounds in an atmospheric compartment did not begin until the late 1970's, when atrazine was found in Maryland rain. Subsequent studies have focused on rain, and one or more triazine herbicides have been detected at high levels in major corn-producing areas, the northcentral, and northeastern United States. Atrazine is a fairly persistent compound and residues in rain have been detected at several locations in the nation throughout the year. Acetanilide herbicides are frequently used in conjunction with triazine herbicides, especially in corn and soybean production. They are not as long-lived as the triazines, but they have been detected in rain at equivalent and even higher concentrations. Many other types of herbicides are used in agriculture, and many of them have been found in the air or precipitation throughout the United States when they have been included as target analytes.

Much of the occurrence of pesticides in the atmosphere can be attributed to agricultural use because of the large acreage involved and the large quantities used, but urban use is also considerable. Together, these two sources contribute to the widespread distribution of a great variety of pesticides in the atmosphere.

Agricultural pesticide inputs to the atmosphere occur during the application process through evaporation and drift, and post-application through volatilization and wind erosion. The degree of losses during and after application depends, in part, on the application method, pesticide formulation, and local meteorology. The airborne concentration of the drifting pesticide is highest near the application site, with the larger drifting droplets and particles dropping out quickly, while atmospheric turbulence diffuses and dilutes the drifting vapors and smaller diameter aerosols. Material that reaches the target, as well as the deposited off-target material, can also volatilize. This volatilization is continuous, even though the rate is variable, and is a significant dissipation route for many pesticides.

Once a pesticide becomes suspended in the atmosphere it will distribute itself between the vapor, aqueous, and particle phases in order to reach an equilibrium condition. The extent of this equilibrium partitioning is dependent on the physical and chemical properties of each pesticide, such as its water solubility and vapor pressure, as well as environmental factors such as temperature, moisture, and the nature and concentration of suspended particulate matter. These factors also dictate the atmospheric lifetime of the pesticide through their effects on the transformation reaction rates and depositional rates to a surface.

Removal of airborne pesticides occurs through wet and dry deposition, and chemical and photochemical reactions. Wet deposition occurs as water vapor condenses around suspended particles as a droplet or snowflake forms and falls to earth, as a raindrop or snowflake physically impacts the particulate material as it falls, and as pesticide vapor partitions into the falling droplet or snowflake. Rain, snow, fog, frost and dew enhance the removal rate of suspended pesticides and clean the atmosphere. Dry deposition is a continuous process that includes both particles and vapors. The drifting vapors and smaller aerosol particles that are not removed from the atmosphere and deposited near the source can be carried, in some cases, great distances downwind. Certain meteorological conditions such as thunderstorms, can move these airborne pesticide vapors and particles into the upper troposphere. Once there, they can become distributed regionally and even globally. The extent of this distribution is dependent upon a variety of factors, including the physical and chemical properties of the pesticide and their transformation reaction rates, in addition to the prevailing meteorology.

Pesticide occurrences in rain, air, and fog often show seasonal trends, with the highest concentrations corresponding to local use and planting seasons. However, pesticides also have been detected during periods before and after the use and planting season. These off-season occurrences could be due to volatilization and wind erosion of previously applied material, or be the result of long-range transport from areas whose planting/growing season started earlier. The most persistent pesticides, such as the organochlorine insecticides DDT, dieldrin, and toxaphene, have been detected in the atmosphere at low levels throughout the year even though they are no longer used in the United States.

Urban pesticide use, which includes consumer applications around the home and professional applications in industrial settings, golf courses, parks, cemeteries, roadways, and railroads, is not as well documented or as well studied as agricultural pesticide use. In home use, the pesticide application rates are specified on the product, but the actual application rates are unregulated and it is difficult to make any meaningful comparisons to agricultural use with the available information. Few published studies have been done that looked for pesticide concentrations in the urban atmosphere, or compared urban pesticide use to agricultural use.

The potential contribution of pesticides from the atmosphere to a surface-water body depends on pesticide levels in atmospheric deposition and on how much of the water budget is derived from surface runoff and direct precipitation. The relative importance of atmospheric inputs to surface waters compared to other nonpoint sources is, generally, proportional to the surface area of the body of water compared to its terrestrial drainage area. The importance of current and past pesticide inputs is less well known largely because of the lack of atmospheric concentration data available for these compounds. Very little research has been done on the depositional inputs of pesticides into surface waters outside the Great Lakes area and even less has been done on pesticides other than organochlorine compounds.

The most clearly documented effects of atmospheric pesticides on human health and aquatic life are related to long-lived, environmentally stable organochlorine insecticides that concentrate in organisms through biomagnification (food chain accumulation), bioconcentration (partitioning), or both. Through these processes, organochlorine insecticides even at the low levels frequently found in air, rain, snow, and fog, have been found to concentrate to significant levels in fish, mammals and humans. Determining the significance of observed pesticide concentrations of non-organochlorine pesticides in air, rain, snow, and fog, is difficult because there are no existing national standards or guidelines for these matrices. The only available guidelines are the U.S. Environmental Protection Agency (USEPA) regulations for contaminants that may occur in public water systems that can adversely affect human health. These include the regulatory maximum contaminant levels (MCLs) and the one-day and long-term health advisories for children. The Occupational Safety and Health Administration (OSHA) also has set limits for airborne pesticide exposure in the workplace in time-weighted average (TWA) concentrations. In addition to human health concerns, aquatic organisms are often more sensitive to low-level pesticide exposures than are humans, and the USEPA and the National Academy of Sciences (NAS) have set water-quality criteria for several pesticides for the protection of aquatic life.

In most cases the measured pesticide concentrations in rain are one or more orders of magnitude below the human health advisory values. There are several instances, though, where the concentrations in rain have exceeded the MCL values in or near agricultural areas. In general, these high rain concentrations occurred infrequently, but they did occur and could occur again. Six pesticides--carbaryl, chlorpyrifos, diazinon, malathion, methidathion, and parathion--were detected in fog and frequently exceeded both USEPA and NAS water-quality criteria for aquatic-life.

Based on the conclusions of this review, pesticides do occur in the atmosphere, some at seasonally high concentrations and others at continuous, very low concentrations. Airborne pesticides and their transformation products are continuously deposited on every surface on the earth as dry deposition and in rain, snow, and fog. Atmospheric deposition, however, is not a one-way process and the deposited chemicals can be reintroduced into the atmosphere for further downwind dispersal. The airborne movement of pesticides in the environment has been described as a global gas chromatographic system where pesticide molecules move many times between the vapor-soil-water-vegetation phases in maintaining an equilibrium of chemical potential (fugacity) between the phases. Atmospheric deposition of pesticides in surface water will not stop when the specific pesticides have stopped being used locally or even nationally. Remaining residues in the soil from past use are continuously introduced into the atmosphere by volatilization and wind erosion, transported downwind, deposited to the surface of the earth, and reintroduced into the atmosphere. This process is repeated continually until the pesticide and transformation products are ultimately mineralized to carbon dioxide and water. Long-range atmospheric transport of organochlorine pesticides from sources outside the United States and Canada also contribute to the national burden. The extent of pesticides in our atmosphere and their deposition into surface waters is not known due to the lack of a consistent, nationwide monitoring network that analyses atmospheric deposition for pesticides and their transformation products. The importance of atmospheric deposition of pesticides into surface waters and its effect on water quality is only beginning to be understood. The importance of atmospheric deposition of pesticides into surface waters and the effect of chronic low-level exposure to a multitude of pesticide residues on long-term human and aquatic health is not known. Much more work needs to be done to fully understand the impact of pesticides in the atmosphere on quality and health of the environment and its inhabitants.

REFERENCES

Addison, R.F., and Smith, T.G., 1974, Organochlorine residue levels in arctic ringed seals: Variation with age and sex: *Oikos*, v. 25, no. 3, p. 335-377.

Addison, R.F., and Zinck, M.E., 1986, PCBs have declined more than DDT-group residues in arctic ringed seals (*Phoca hispida*) between 1972 and 1981: *Environ. Sci. Technol.*, v. 20, no. 3, p. 253-256.

Agarwal, H.C., Kaushik, C.P., and Pillai, M.K.K., 1987, Organochlorine insecticide residues in the rain water in Delhi, India: *J. Water, Air, Soil Poll.*, v. 32, p. 293-302.

Akesson, N.B., and Yates, W.E., 1964, Problems relating to application of agricultural chemicals and resulting drift residues, *in* Smith, R.F., and Mittler, T.E., eds., *Annual review of entomology*: Annual Reviews, Inc., Palo Alto, California, p. 285-318.

Amundsen, C.E., Hanssen, J.E., Semb, A., and Steinnes, E., 1992, Long-range atmospheric transport of trace elements to southern Norway: *Atmospheric Environ.*, v. 26A, no. 7, p. 1309-1324.

Andrilenas, P.A., 1974, Farmers' use of pesticides in 1971--Quantities: U. S. Department of Agriculture, Economic Research Service, National Economics Analysis Division, Agricultural Economic Report 252, 56 p.

Anspaugh, L.R., Shinn, J.H., Phelps, P.L., and Kennedy, N.C., 1975, Resuspension and redistribution of plutonium in soils: *Health Physics*, v. 29, p. 571-582.

Antommaria, P., Corn, M., and DeMaio, L., 1965, Airborne particulates in Pittsburgh: Association with *p,p'*-DDT: *Science*, v. 150, p. 1476-1477.

Appel, B.R., Tokiwa, Y., and Kothny, E.L., 1983, Sampling the carbonaceous particles in the atmosphere: *Atmospheric Environ.*, v. 17, no. 9, p. 1787-1796.

Arimoto, R., 1989, Atmospheric deposition of chemical contaminants to the Great Lakes: *J. Great Lakes Research*, v. 15, no. 2, p. 339-356.

Arthur, R.D., Cain, J.D., and Barrentine, B.F., 1976, Atmospheric levels of pesticides in the Mississippi delta: *Bull. Environ. Contam. Toxicol.*, v. 15, no. 2, p. 129-134.

Aspelin, A.L., 1994, Pesticides industry sales and usage; 1992 and 1993 market estimates: U.S. Environmental Protection Agency, Washington, D.C., 733-K-94-001, 33 p.

Aspelin, A.L., Grube, A.H., and Torla, R., 1992, Pesticides industry sales and usage; 1990 and 1991 market estimates: U.S. Environmental Protection Agency, Washington, D.C., 733-K-92-001, 35 p.

Atkinson, R., 1989, Kinetics and mechanisms of the gas-phase reactions of the hydroxyl radical with organic compounds: *J. Phys. Chem. Ref. Data*, Monograph no. 1, p. 246.

Atkinson, R., and Carter, W.P.L., 1984, Kinetics and mechanisms of the gas-phase reactions of ozone with organic compounds under atmospheric conditions: *Chem. Rev.*, v. 84, no. 5, p. 437-470.

Atlas, E., and Giam, C.S., 1988, Ambient concentration and precipitation scavenging of atmospheric organic pollutants: *J. Water, Air, Soil Poll.*, v. 38, p. 19-36.

Atlas, E.L., and Schauffler, S., 1990, Concentration and variation of trace organic compounds in the North Pacific atmosphere, *in* Kurtz, D.A., ed., *Long range transport of pesticides*: Lewis Publishers, Inc., Chelsea, MI, p. 161-184.

Ballschmiter, K., and Wittlinger, R., 1991, Interhemisphere exchange of hexachloro-cyclohexanes, hexachlorobenzene, polychlorobiphenyls, and 1,1,1-trichloro-2,2-bis(p-chlorophenyl)ethane in the lower troposphere: *Environ. Sci. Technol.*, v. 25, no. 6, p. 1103-1111.

Bardsley, C.E., Savage, K.E., and Walker, J.C., 1968, Trifluralin behavior in soil. II. Volatilization as influenced by concentration, time, soil moisture content, and placement: *Agronomy J.*, v. 60, p. 89-92.

Barnes, C.J., Lavy, T.L., and Mattice, J.D., 1987, Exposure of non-applicator personnel and adjacent areas to aerially applied propanil: *Bull. Environ. Contam. Toxicol.*, v. 39, p. 126-133.

Barrie, L.A., 1986, Arctic air pollution--An overview of current knowledge: *Atmospheric Environ.*, v. 20, no. 4, p. 643-663.

Barrie, L.A., and Schemenauer, R.S., 1986, Pollutant wet deposition mechanisms in precipitation and fog water: *J. Water, Air, Soil Poll.*, v. 30, p. 91-104.

Batchelor, G.S., and Walker, K.C., 1954, Health hazards involved in use of parathion in fruit orchards on north central Washington: *A.M.A. Archives of Industrial Hygiene and Occupational Medicine*, v. 10, p. 522-529.

Behymer, T.D., and Hites, R.A., 1985, Photolysis of polycyclic aromatic hydrocarbons adsorbed on simulated atmospheric particulates: *Environ. Sci. Technol.*, v. 19, no. 10, p. 1004-1006.

Bevenue, A., Ogata, J.N., and Hylin, J.W., 1972, Organochlorine pesticides in rainwater, Oahu, Hawaii, 1971-1972: *Bull. Environ. Contam. Toxicol.*, v. 8, no. 4, p. 238-241.

Bidleman, T.F., 1988, Atmospheric processes: *Environ. Sci. Technol.*, v. 22, no. 4, p. 361-367.

Bidleman, T.F., Billings, W.N., and Foreman, W.T., 1986, Vapor-particle partitioning of semivolatile organic compounds--Estimates from field collections: *Environ. Sci. Technol.*, v. 20, no. 10, p. 1038-1043.

Bidleman, T.F., and Christensen, E.J., 1979, Atmospheric removal processes for high molecular weight organochlorines: *J. Geophys. Res.*, v. 84, no. C12, p. 7857-7862.

Bidleman, T.F., and Foreman, W.T., 1987, Vapor-particle partitioning of semivolatile organic compounds, *in* Hites, R.A., and Eisenreich, S.J., eds., *Sources and fates of aquatic pollutants*: American Chemical Society, Washington, D.C., p. 27-56.

Bidleman, T.F., and Leonard, R., 1982, Aerial transportation of pesticides over the northern Indian Ocean and adjacent seas: *Atmospheric Environ.*, v. 16, no. 5, p. 1099-1107.

Bidleman, T.F., and Olney, C.E., 1974, High volume collection of atmospheric polychlorinated biphenyls: *Bull. Environ. Contam. Toxicol.*, v. 11, p. 442-450.

Bidleman, T.F., Patton, G.W., Hinckley, D.A., Walla, M.D., Cotham, W.E., and Hargrave, B.T., 1990, Chlorinated pesticides and polychlorinated biphenyls in the atmosphere of the Canadian arctic, *in* Kurtz, D.A., ed., *Long range transport of pesticides*: Lewis Publishers, Inc., Chelsea, MI, p. 237-372.

Bidleman, T.F., Patton, G.W., Walla, M.D., Hargrave, B.T., Vass, W.P., Erickson, P., Fowler, B., Scott, V., and Gregor, D.J., 1989, Toxaphene and other organochlorines in arctic ocean fauna--Evidence for atmospheric delivery: *Arctic*, v. 42, no. 4, p. 307-313.

Bidleman, T.F., Wideqvist, U., Jansson, B., and Söderlund, R., 1987, Organochlorine pesticides and polychlorinated biphenyls in the atmosphere of southern Sweden: *Atmospheric Environ.*, v. 21, no. 3, p. 641-654.

Bidleman, T.F., Zaranski, M.T., and Walla, M.D., 1988, Toxaphene: Usage, aerial transport and deposition, *in* Schmidtke, N.W., ed., *Toxic contamination in large lakes*: Lewis Publishers, Inc., Chelsea, MI, p. 257-284.

Bigelow, D.S., 1984, NADP/NTN Site selection and installation: National Atmospheric Deposition Program, Instruction Manual Report, p. 23.

Billings, W.N., and Bidleman, T.F., 1983, High volume collection of chlorinated hydrocarbons in urban air using three solid adsorbents: *Atmospheric Environ.*, v. 17, no. 2, p. 383-391.

Björseth, A., and Lunde, G., 1979, Long-range transport of polycyclic aromatic hydrocarbons: *Atmospheric Environ.*, v. 13, no. 1, p. 45-53.

Brooksbank, P., 1983, The Canadian network for sampling organic compounds in precipitation: Technical bulletin No. 129, Environment Canada, Water Quality Branch, Ottawa, Canada.

Brown, M.A., Petreas, M.X., Okamoto, H.S., Mischke, T.M., and Stephens, R.D., 1993, Monitoring of malathion and its impurities and environmental transformation products on surfaces and in air following an aerial application: *Environ. Sci. Technol.*, v. 27, no. 2, p. 388-397.

Brun, G.L., Howell, G.D., and O'Neill, H.J., 1991, Spatial and temporal patterns of organic contaminants in wet precipitation in Atlantic Canada: *Environ. Sci. Technol.*, v. 25, no. 7, p. 1249-1261.

Burgoyne, T.W., and Hites, R.A., 1993, Effects of temperature and wind direction on the atmospheric concentrations of alpha-endosulfan: *Environ. Sci. Technol.*, v. 27, no. 6, p. 910-914.

Businger, J.A., 1986, Evaluation of the accuracy with which dry deposition can be measured with current micrometeorological techniques: *J. Climate and Appl. Meteorol.*, v. 25, no. 8, p. 1100-1124.

Cade, T.J., White, C.M., and Haugh, J.R., 1968, Peregrines and pesticides in Alaska: *The Condor*, v. 70, no. 2, p. 170-178.

California Department of Food and Agriculture, 1984-1986, Pesticide use enforcement: California Department of Food and Agriculture, Chemical Residues Report.

California Department of Pesticide Regulations, 1990, Pesticide use data: Computer tapes available from California Department of Pesticide Regulations, Sacramento, California.

Capel, P.D., 1991, Wet atmospheric deposition of herbicides in Minnesota: *in* Proc. of the technical meeting of the U.S. Geological Survey toxic substances hydrology program: U.S. Geological Survey Water Resources Investigations Report 91-4034, p. 334-337.

Capel, P.D., Gunde, R., Zurcher, F., and Giger, W., 1990, Carbon speciation and surface tension of fog: *Environ. Sci. Technol.*, v. 24, no. 5, p. 722-727.

Capel, P.D., Leuenberger, C., and Giger, W., 1991, Hydrophobic organic chemicals in urban fog: *Atmospheric Environ.*, v. 25A, no. 7, p. 1335-1346.

Caplan, P.E., Culver, D., and Theilen, W.C., 1956, Human exposure in populated areas during airplane application of malathion: *A.M.A. Arch. of Ind. Health*, v. 14, p. 326-332.

Caro, J.H., Taylor, A.W., and Lemon, E.R., 1971, Measurement of pesticide concentrations in the air overlying a treated field: *Proc. of the international symposium on identification and measurement of environmental pollutants*, 14-17 June, Ottawa, Ontario, Canada, p. 72-77.

Cessna, A.J., and Muir, D.C.G., 1991, Photochemical transformations, *in* Grover, R., and Cessna, A.J., eds., *Environmental chemistry of herbicides*. V. 2: CRC Press, Boca Raton, FL, p. 199-264.

Chan, C.H., and Perkins, L.H., 1989, Monitoring of trace organic contaminants in atmospheric precipitation: *J. Great Lakes Res.*, v. 15, no. 3, p. 465-475.

Chang, J.S., and Penner, J.E., 1978, Analysis of global budget of halocarbons: *Atmospheric Environ.*, v. 12, p. 1867-1873.

Chang, L.W., Atlas, E., and Giam, C.S., 1985, Chromatographic separation and analysis of chlorinated hydrocarbons and phthalic acid esters from ambient air samples: *Int. J. Environ. Anal. Chem.*, v. 19, p. 145-153.

Chepil, W.S., 1945, Dynamics of wind erosion, III: *Soil Sci.*, v. 60, p. 475-480.

Chepil, W.S., and Woodruff, N.P., 1963, The physics of wind erosion and its control, *in* Norman, A.G., ed., *Advances in agronomy*: Academic Press, New York, p. 211-302.

Chiou, C.T., and Shoup, T.D., 1985, Soil sorption of organic vapors and effects of humidity on sorptive mechanism and capacity: *Environ. Sci. Technol.*, v. 19, no. 12, p. 1196-1200.

Christensen, E.J., Olney, C.E., and Bidleman, T.F., 1979, Comparison of dry and wet surfaces for collecting organochlorine dry deposition: *Bull. Environ. Contam. Toxicol.*, v. 23, p. 196-202.

Christensen, P., Yates, W.E., and Akesson, N.B., 1969, Meteorology and drift: Presented at the Fourth International Agricultural Aviation Congress research apparatus and techniques, chemicals and distribution, 25-29 August, Kingston, Ontario, Canada.

Chukwudebe, A., March, R.B., Othman, M., and Fukuto, T.R., 1989, Formation of trialkyl phosphorothioate esters from organophosphorous insecticides after exposure to either ultraviolet light or sunlight: *J. Agri. Food Chem.*, v. 37, no. 2, p. 539-545.

Chyou, S.W., and Sleicher, C.A., 1986, Vaporization and dispersion from a surface to a turbulent boundary layer: *Ind. Eng. Chem. Fundam.*, v. 25, no. 4, p. 659-661.

Clendening, L.D., Jury, W.A., and Ernst, F.F., 1990, A field mass balance study of pesticide volatilization, leaching, and persistence, *in* Kurtz, D.A., ed., *Long range transport of pesticides*: Lewis Publishers, Inc., Chelsea, MI, p. 47-60.

Cliath, M.M., Spencer, W.F., Farmer, W.J., Shoup, T.D., and Grover, R., 1980, Volatilization of S-ethyl N,N-dipropyl-thiocarbamate from water and wet soil during and after flood irrigation of an alfalfa field: *J. Agri. Food Chem.*, v. 28, no. 3, p. 610-613.

Cohen, J.M., and Pinkerton, C., 1966, Widespread translocation of pesticides by air transport and rain-out, *in* Rosen, A.A., and Kraybill, H.F., eds., *Organic pesticides in the environment*: American Chemical Society, Washington, D.C., p. 163-176.

Connors, V.S., Miles, T., and Reichle, H.G., Jr., 1989, Large-scale transport of a CO-enhanced air mass from Europe to the Middle East: *J. Atmospheric Chem.*, v. 9, p. 479-496.

Cotham, W.E., and Bidleman, T.F., 1991, Estimating the atmospheric deposition of organochlorine contaminants to the Arctic: *Chemosphere*, v. 22, no. 1-2, p. 165-188.

Cotham, W.E., and Bidleman, T.F., 1992, Laboratory investigations of the partitioning of organochlorine compounds between the gas phase and atmospheric aerosols on glass fiber filters: *Environ. Sci. Technol.*, v. 26, no. 3, p. 469-478.

Coutant, R.W., Brown, L., Chang, J.C., Riggin, R.M., and Lewis, R.G., 1988, Phase distribution and artifact formation in ambient air sampling for polynuclear aromatic hydrocarbons: *Atmospheric Environ.*, v. 22, no. 2, p. 403-409.

Coutant, R.W., Callahan, P.J., and Kuhlman, M.R., 1989, Design and performance of a high-volume compound annular denuder: *Atmospheric Environ.*, v. 23, no. 10, p. 2205-2211.

Coutts, H.H., and Yates, W.E., 1968, Analysis of spray droplet distributions from agricultural aircraft: *Trans. American Soc. Agri. Eng.*, v. 11, p. 25-27.

Crosby, D.G., 1976, Herbicide photodecomposition, *in* Kearney, P.C., and Kaufman, D.D., eds., *Herbicides-chemistry, degradation, and mode of action*: Marcel Dekker, Inc., New York, p. 835-890.

Crosby, D.G., and Li, M.Y., 1969, Herbicide photodecomposition, *in* Kearney, P.C., and Kaufman, D.D., eds., *Degradation of herbicides*: Marcel Dekker, Inc., New York, p. 321-363.

Crosby, D.G., Li, M.Y., Seiber, J.N., and Winterlin, W.L., 1981, Environmental monitoring of MCPA in relation to orchard contamination: California Department of Food and Agriculture, EH-88-9, 146 p.

Crosby, D.G., and Moilanen, K.W., 1974, Vapor-phase photodecomposition of aldrin and dieldrin: *Arch. Environ. Contam. Toxicol.*, v. 2, p. 62.

-----1977, Vapor-phase photodecomposition of DDT: *Chemosphere*, v. 6, p. 167.

Culver, D., Caplan, P.E., and Batchelor, G.S., 1956, Studies of human exposure during aerosol application of malathion and chlorthion: *A.M.A. Arch. Indust. Health*, v. 13, p. 37-50.

Cunningham, R.T., Brann, J.L., Jr., and Fleming, G.A., 1962, Factors affecting the evaporation of water from droplets in airblast spraying: *J. Econ. Entomol.*, v. 55, no. 2, p. 192-199.

Czeplak, G., and Junge, C., 1974, Studies of interhemispheric exchange in the troposphere by a diffusion model, *in* Frenkiel, F.N., and Munn, R.E., eds., *Turbulent diffusion in environmental pollution*: Academic Press, New York, p. 57-72.

Daines, R.H., 1952, 2,4-D as an air pollutant and its effects on various species of plants, *in* McCabe, L.C., ed., Air Pollution, *Proc. of the U.S. Tech. Conf. on Air Pollution*: McGraw-Hill Book Co., Inc., New York, p. 140-143.

Dasch, J.M., 1985, Direct measurement of dry deposition to a polyethylene bucket and various surrogate surfaces: *Environ. Sci. Technol.*, v. 19, p. 721-725.

Davidson, C.I., 1989, Mechanisms of wet and dry deposition of atmospheric contaminants to snow surfaces, *in* Oeschger, H., and Langway, C.C.L., Jr., eds., *The environmental record in glaciers and ice sheets:* John Wiley and Sons, New York, p. 29-51.

Dickerson, R.R., Huffman, G.J., Luke, W.T., Nunnermacker, L.J., Pickering, K.E., Leslie, A.C.D., Linsey, C.G., Slinn, W.G.N., Kelly, T.J., Daum, P.H., Delany, A.C., Greenberg, J.P., Zimmermann, P.R., Boatman, J.F., Ray, J.D., and Stedman, D.H., 1987, Thunderstorms--An important mechanism in the transport of air pollutants: *Science*, v. 235, p. 460-465.

Droppo, J.G., Jr., 1985, Concurrent measurements of ozone dry deposition using eddy correlation and profile flux methods: *J. Geophys. Res.*, v. 90, no. D1, p. 2111-2118.

Edwards, C.A., 1986, Agrochemicals as environmental pollutants, *in* Benjt, V.H., ed., *Control of pesticide application and residues in food--A guide and directory*: Swedish Scientific Press, Uppsala, Sweden, Geo. Ekstrom., p. 1-20.

Eisenreich, S.J., 1987, The chemical limnology of nonpolar organic contaminants: Polychlorinated biphenyls in Lake Superior, *in* Eisenreich, S.J., and Hites, R.A., eds., *Sources and fates of aquatic pollutants*: American Chemical Association, Washington, D.C., p. 393-469.

Eisenreich, S.J., Bidleman, T.F., Murphy, T.J., Davis, A.R., Banning, D.A., Giam, C.S., Priznar, F.J., and Mullin, M.D., 1980, Trace organics-a review and assessment, *in* Galloway, J.N., Eisenreich, S.J., and Scott, B.C., eds., *Toxic substances in atmospheric deposition--A review and assessment:* National Atmospheric Deposition Program, NC-141, p. 83-113.

Eisenreich, S.J., Looney, B.B., and Thornton, J.D., 1981, Airborne organic contaminants in the Great Lakes ecosystem: *Environ. Sci. Technol.*, v. 15, no. 1, p. 30-38.

Ernst, W.R., Doe, P.J., Hennigar, J., and Hennigar, P., 1991, Toxicity to aquatic organisms of off-target deposition of endosulfan applied by aircraft: *Environ. Toxicol. Chem.*, v. 10, no. 1, p. 103-114.

Farm Chemicals, 1992, Chemical abuse by nation's homeowners?: *Farm Chemicals*, v. 155, no. 8, p. 10.

Farwell, S.O., Robinson, E., Powell, W.J., and Adams, D.F., 1976, Survey of airborne 2,4-D in south-central Washington: *J. Air Poll. Control Assoc.*, v. 26, no. 3, p. 224-230.

Finkelstein, H., 1969, Air pollution aspects of pesticides: Washington, D.C., National Air Pollution Control Administration, NTIS PB-188091, 169 p.

Finlayson-Pitts, B.J., and Pitts, J.N., 1986, *Atmospheric chemistry: Fundamentals and experimental techniques*: John Wiley and Sons, Inc., New York, 1,098 p.

Fleck, J., Ross, L., Tran, D., Melvin, J., and Fong, B., 1991, Off-target movement of endosulfan from artichoke fields in Monterey county: California Department of Food and Agriculture, EH 91-5, 17 p.

Foreman, W.T., and Bidleman, T.F., 1987, An experimental system for investigating vapor-particle partitioning of trace organic pollutants: *Environ. Sci. Technol.*, v. 21, no. 9, p. 869-875.

-----1990, Semivolatile organic compounds in the ambient air of Denver, CO: *Atmospheric Environ.*, v. 24A, no. 9, p. 2405-2416.

Foster, J.R., Pribush, R.A., and Carter, B.H., 1990, The chemistry of dews and frosts in Indianapolis, Indiana: *Atmospheric Environ.*, v. 24A, p. 2229-2236.

Frank, R., Johnson, K., Braun, H.E., Halliday, C.G., and Harvey, J., 1991, Monitoring air, soil, stream and fish for aerial drift of permethrin: *Environ. Monitor. Assess.*, v. 16, p. 137-150.

Franz, T.P., Eisenreich, S.J., and Swanson, M.B., 1991, Evaluation of precipitation samplers for assessing atmospheric fluxes of trace organic contaminants: *Chemosphere*, v. 23, no. 3, p. 343-362.

Frost, K.R., and Ware, G.W., 1970, Pesticide drift from aerial and ground applications: *Agri. Eng.*, v. 51, no. 8, p. 460-467.

George, J.L., and Frear, E.H., 1966, Pesticides in Antarctica: *J. Applied Ecol.*, v. 3, no. 3, p. 155-176.

Giam, C.S., Atlas, E., Chan, H.S., and Neff, G.S., 1980, Phthalate esters, PCB and DDT residues in the Gulf of Mexico atmosphere: *Atmospheric Environ.*, v. 14, p. 65-69.

Gianessi, L.P., and Puffer, C.A., 1990 [1991], Herbicide use in the United States: *Resour. for the Future*, December 1990, 128 p.

-----1992a, Fungicide use in U.S crop production: *Resour. for the Future*, November 1992.

-----1992b, Insecticide use in U.S. crop production: *Resour. for the Future*, August 1992.

Giddings, W.P., and Baker, M.B., 1977, Sources and effects of monolayers on atmospheric water droplets: *J. Atmospheric Sci.*, v. 34, no. 12, p. 1957-1964.

Gill, P.S., Graedel, T.E., and Weschler, C.J., 1983, Organic films of atmospheric aerosol particles, fog droplets, cloud droplets, raindrops, and snowflakes: *Rev. of Geophys. Space Phys.*, v. 21, no. 4, p. 903-920.

Gillette, D.A., 1983, Threshold velocities for wind erosion on natural terrestrial arid surfaces (a summary), *in* Pruppacher, H.R., Semonin, R.G., and Slinn, W.G.N., eds., *Precipitation scavenging, dry deposition, and resuspension*: Elsevier, Amsterdam, v. 2, p. 1047-1058.

Gilliom, R.J., Alley, W.M., and Gurtz, M.E., 1995, Design of the National Water-Quality Assessment Program: Occurrence and distribution of water-quality conditions: U.S. Geological Survey Circular 1112, 33 p.

Gilliom, R.J., Alexander, R.B., and Smith, R.A., 1985, Pesticides in the nation's rivers, 1975-1980, and implications for future monitoring: U.S. Geological Survey Water-Supply Paper 2271, 26 p.

Glotfelty, D.E., 1978, The atmosphere as a sink for applied pesticides: *J. Air Poll. Control Assoc.*, v. 28, no. 9, p. 917-921.

-----1981, Atmospheric dispersion of pesticides from treated fields: College Park, University of Maryland, Ph.D. dissertation, 512 p.

-----1987, The effects of conservation tillage practices on pesticides volatilization and degradation, *in* Logan, T.J., Davidson, J.M., Baker, J.L., and Overcash, M.R., eds., *Effects of conservation tillage on groundwater quality-nitrates and pesticides*: Lewis Publishers, Chelsea, MI, p. 169-177.

Glotfelty, D.E., and Caro, J.H., 1975, Introduction, transport, and fate of persistent pesticides in the atmosphere, *in* Deitz, V.R., ed., *Removal of trace contaminants from the air*: American Chemical Society, Washington, D.C., p. 42-62.

Glotfelty, D.E., Leech, M.M., Jersey, J., and Taylor, A.W., 1989, Volatilization and wind erosion of soil surface applied atrazine, simazine, alachlor, and toxaphene: *J. Agri. Food Chem.*, v. 37, no. 2, p. 546-551.

Glotfelty, D.E., Majewski, M.S., and Seiber, J.N., 1990a, Distribution of several organophosphorous insecticides and their oxygen analogs in a foggy atmosphere: *Environ. Sci. Technol.*, v. 24, no. 3, p. 353-357.

Glotfelty, D.E., Schomburg, C.J., McChesney, M.M., Segebiel, J.C., and Seiber, J.N., 1990b, Studies of the distribution, drift, and volatilization of diazinon resulting from spray application to a dormant peach orchard: *Chemosphere*, v. 21, no. 10-11, p. 1303-1314.

Glotfelty, D.E., Seiber, J.N., and Liljedahl, L.A., 1987, Pesticides in fog: *Nature*, v. 325, p. 602-605.

Glotfelty, D.E., Taylor, A.W., Turner, B.C., and Zoller, W.H., 1984, Volatilization of surface-applied pesticides from fallow soil: *J. Agri. Food Chem.*, v. 32, no. 3, p. 638-643.

Glotfelty, D.E., Williams, G.H., Freeman, H.P., and Leech, M.M., 1990c, Regional atmospheric transport and deposition of pesticides in Maryland, *in* Kurtz, D.A., ed., *Long-range transport of pesticides*: Lewis Publishers, Inc., Chelsea, MI, p. 199-222.

Goering, C.E., and Butler, B.J., 1975, Paired field studies of herbicide drift: *Transactions of the American Society of Agricultural Engineers*, v. 18, p. 27-34.

Gold, A.J., and Groffman, P.M., 1993, Leaching of agrichemicals from suburban areas, *in* Racke, K.D., and Leslie, A.R., eds., *Pesticides in urban environments*: American Chemical Society, Washington, D.C., p. 183-190.

Goldberg, E.D., 1975, Synthetic organohalides in the sea: *Proc. of the Royal Society of London*, v. 189, p. 277-289.

Goolsby, D.A., Thurman, E.M., Pommes, M.L., and Battaglin, W.A., 1994, Temporal and geographic distribution of herbicides in precipitation in the midwest and northeast United States, 1990-91: *in* Weigmann, D.L., ed., New directions in pesticide research, development, management, and policy, *Proc. of the Fourth National Pesticide Conference*, Richmond, Virginia, Nov. 1-3, 1993.

Goss, K.U., 1993, Effects of temperature and relative humidity on the sorption of organic vapors on clay minerals, *Environ. Sci. Technol.*, v. 27, no. 10, p. 2127-2132.

Graedel, T.E., Gill, P.S., and Weschler, C.J., 1983, Effects of organic surface films on the scavenging of atmospheric gases by raindrops and aerosol particles, *in* Pruppacher, H.R., Semonin, R.G., and Slinn, W.G.N., eds., *Precipitation scavenging, dry deposition, and resuspension*: Elsevier, New York, v. 1, p. 417-430.

Gregor, D.J., 1990, Deposition and accumulation of selected agricultural pesticides in Canadian arctic snow, *in* Kurtz, D.A., ed., *Long range transport of pesticides*: Lewis Publishers, Inc., Chelsea, MI, p. 373-386.

Gregor, D.J., and Gummer, W.D., 1989, Evidence of atmospheric transport and deposition of organochlorine pesticides and polychlorinated biphenyls in Canadian arctic snow: *Environ. Sci. Technol.*, v. 23, no. 5, p. 561-565.

Grosjean, D., 1983, Polycyclic aromatic hydrocarbons in Los Angeles air from samples collected on Teflon, glass and quartz filters: *Atmospheric Environ.*, v. 17, no. 12, p. 2565-2573.

Grover, R., 1974, Herbicide entry into the atmospheric environment: *Chem. in Canada*, Summer, p. 36-38.

-----1989, Magnitude and source of airborne residues of herbicides in Saskatchewan, *in* Dosman, J.A., and Cockcroft, D.W., eds., *Principles of health and safety in agriculture*: CRC Press, Inc., Boca Raton, FL, p. 222-225.

-----1991, Nature, transport, and fate of airborne residues, *in* Grover, R., and Cessna, A.J., eds., *Environmental chemistry of herbicides*: CRC Press, Boca Raton, FL, p. 89-118.

Grover, R., Kerr, L.A., Bowren, K.E., and Khan, S.U., 1988a, Airborne residues of triallate and trifluralin in Saskatchewan: *Bull. Environ. Contam. Toxicol.*, v. 40, p. 683-688.

Grover, R., Kerr, L.A., and Khan, S.U., 1981, Multidetector gas chromatographic determination and confirmation of airborne triallate residues in Saskatchewan: *J. Agri. Food Chem.*, v. 29, p. 1082-1084.

Grover, R., Kerr, L.A., Maybank, J., and Yoshida, K., 1978, Field measurement of droplet drift from ground sprayers. I. Sampling, analytical and data integration techniques: *Can. J. Plant Sci.*, v. 58, p. 611-622.

Grover, R., Kerr, L.A., Wallace, K., Yoshida, K., and Maybank, J., 1976, Residues of 2,4-D in air samples from Saskatchewan-1966-1975: *J. Environ. Sci. Health*, v. B11, no. 4, p. 331-347.

Grover, R., Maybank, J., and Yoshida, K., 1972, Droplet and vapor drift from butyl ester and dimethylamine salt of 2,4-D: *Weed Sci.*, v. 20, no. 4, p. 320-324.

Grover, R., Shewchuk, S.R., Cessna, A.J., Smith, A.E., and Hunter, J.H., 1985, Fate of 2,4-D iso-octyl ester after application to a wheat field: *J. Environ. Qual.*, v. 14, no. 2, p. 203-210.

Grover, R., Smith, A.E., Shewchuk, S.R., Cessna, A.J., and Hunter, J.H., 1988b, Fate of trifluralin and triallate applied as a mixture to a wheat field: *J. Environ. Qual.*, v. 17, no. 4, p. 543-550.

Harder, H.W., Christensen, E.C., Matthews, J.R., and Bidleman, T.F., 1980, Rainfall input of toxaphene to a South Carolina estuary: *Estuaries*, v. 3, no. 2, p. 142-147.

Hargrave, B.T., Vass, W.P., Erickson, P.E., and Fowler, B.R., 1988, Atmospheric transport of organochlorines to the Arctic Ocean: *Tellus*, v. 40B, p. 480-493.

Harper, L.A., McDowell, L.L., Willis, G.H., Smith, S., and Southwick, L.M., 1983, Microclimate effects on toxaphene and DDT volatilization from cotton plants: *Agronomy J.*, v. 75, no. Mar-Apr, p. 295-302.

Harper, L.A., White, A.W., Jr., Bruce, R.R., Thomas, A.W., and Leonard, R.A., 1976, Soil and microclimate effects on trifluralin volatilization: *J. Environ. Qual.*, v. 5, no. 3, p. 236-242.

Hart, K.M., and Pankow, J.F., 1994, High-volume air sampler for particle and gas sampling. 2. Use of backup filters to correct for the adsorption of gas-phase polycyclic aromatic hydrocarbons to the front filter: *Environ. Sci. Tech.*, v. 28, no. 4, p. 655-661.

Hart, K.M., Isabelle, L.M., and Pankow, J.F., 1992, High-volume air sampler for particle and gas sampling. 1. Design and gas sampling performance: *Environ. Sci. Technol.*, v. 26, no. 5, p. 1048-1052.

Hartley, G.S., 1969, Evaporation of pesticides, *in* Gould, R.F., ed., *Pesticidal formulations research, physical and colloidal chemical aspects*: American Chemical Society, Washington D.C., p. 115-134.

Harvey, G.R., and Steinhauer, W.G., 1974, Atmospheric transport of polychlorobiphenyls to the North Atlantic: *Atmospheric Environ.*, v. 8, p. 777-782.

Himel, C., Loats, H., and Bailey, G.W., 1990, Pesticide sources to the soil and principles of spray physics, *in* Cheng, H.H., ed., *Pesticides in the soil environment*: Soil Science Society of America, Inc., Madison, WI, p. 7-50.

Hirsch, R.M., Alley, W.M., and Wilber, W.G., 1988, Concepts for a National Water-Quality Assessment Program: U.S. Geological Survey Circular 1021, 42 p.

Hodge, J.E., 1993, Pesticide trends in the professional and consumer market, *in* Racke, K.D., and Leslie, A.R., eds., *Pesticides in urban environments*: American Chemical Society, Washington, D.C., p. 10-17.

Hoff, R.H., Muir, D.C.G., and Grift, N.P., 1992, Annual cycle of polychlorinated biphenyls and organohalogen pesticides in air in southern Ontario. 1. Air concentration data: *Environ. Sci. Technol.*, v. 26, no. 2, p. 266-275.

Högström, U., 1979, Initial dry deposition and type of source in relation to long distance transport of air pollutants: *Atmospheric Environ.*, v. 13, p. 295-301.

Hollingsworth, E.B., 1980, Volatility of trifluralin from field soil: *Weed Sci.*, v. 28, no. 2, p. 224-228.

Holsen, T.M., and Noll, K.E., 1992, Dry deposition of atmospheric particles: Application of current models to ambient data: *Environ. Sci. Technol.*, v. 26, no. 9, p. 1807-1815.

Holton, J.R., 1979, *An introduction to dynamic meteorology*: Academic Press, New York, 391 p.

Howard, P.H., and others, 1991, *Handbook of environmental fate and exposure data for organic chemicals, V. III. Pesticides*: Lewis Publishers, Inc., Chelsea, MI, 684 p.

Hurst, H.R., 1982, Cotton (*Gossypium hirsutum*) response to simulated drift from selected herbicides: *Weed Sci.*, v. 30, p. 311-315.

Iwata, H., Tanabe, S., Sakai, N., and Tatsukawa, R., 1993, Distribution of persistent organochlorines in the ocean air and surface seawater and the role of ocean on their global transport and fate: *Environ. Sci. Technol.*, v. 27, no. 6, p. 1080-1098.

Jacob, D.J., Flagan, R.C., Waldman, J.M., and Hoffman, M.R., 1983, Design and calibration of a rotating arm collector for ambient fog sampling, *in* Pruppacher, H.R., Semonin, R.G., and Slinn, W.G.N., eds., *Precipitation scavenging, dry deposition, and resuspension*: Elsevier, New York., v. 1, p. 125-136.

Jacob, D.J., Waldman, J.M., Haghi, M., Hoffmann, M.R., and Flagan, R.C., 1985, Instrument to collect fogwater for chemical analysis: *Rev. of Sci. Instrum.*, v. 56, no. 6, p. 1291-1293.

Jacoby, P.W., Meadors, C.H., and Clark, L.E., 1990, Effects of triclopyr, clopyralid, and picloram on growth and production of cotton: *J. Prod. Agri.*, v. 3, p. 297-301.

Johnson, N.D., Lane, D.A., Schroeder, W.H., and Strachan, W.M., 1990, Measurements of selected organochlorine compounds in air near Ontario Lakes--Gas-particle relationships, *in* Kurtz, D.A., ed., *Long range transport of pesticides*: Lewis Publishers, Inc., Chelsea, MI, p. 105-114.

Judeikis, H.S., and Siegel, S., 1973, Particle-catalyzed oxidation of atmospheric pollutants: *Atmospheric Environ.*, v. 7, p. 619-631.

Junge, C.E., 1975, Transport mechanisms for pesticides in the atmosphere: *Pure and Appl. Chem.*, v. 42, p. 95-104.

-----1977, Basic considerations about trace constituents in the atmosphere as related to the fate of global pollutants, *in* Suffet, I.H., ed., *Fate of pollutants in the air and water environments*: John Wiley and Sons, New York, p. 7-25.

Kaushik, C.P., Pillai, M.K.K., Raman, A., and Agarwal, H.C., 1987, Organochlorine insecticide residues in air in Delhi, India: *J. Water, Air, Soil Poll.*, v. 32, p. 63-76.

Keller, C.D., and Bidleman, T.F., 1984, Collection of airborne polycyclic aromatic hydrocarbons and other organics with a glass fiber filter-polyurethane foam system: *Atmospheric Environ.*, v. 18, no. 4, p. 837-845.

Kempny, J., Kotzias, D., and Korte, F., 1981, Reaction of atrazine in an adsorbed phase with UV radiation: *Chemosphere*, v. 10, no. 5, p. 487-490.

Klisenko, M.A., and Pis'mennaya, M.V., 1979, *Photochemical conversion of organophosphorous pesticides in air*: Gig. Tr. Prof. Zabol., v. 56, (Russian): through Chem Abst. 94. 126596h (1979).

Knap, A.H., and Binkley, K.S., 1991, Chlorinated organic compounds in the troposphere over the Western North Atlantic Ocean measured by aircraft: *Atmospheric Environ.*, v. 25A, no. 8, p. 1507-1516.

Krieger, M.S., and Hites, R.A., 1992, Diffusion denuder for the collection of semivolatile compounds: *Environ. Sci. Technol.*, v. 26, no. 8, p. 1551-1555.

Kurtz, D.A., and Atlas, E.L., 1990, Distribution of hexachlorocyclohexanes in the Pacific Ocean basin, air and water, 1987, *in* Kurtz, D.A., ed., *Long range transport of pesticides*: Lewis Publishers, Inc., Chelsea, MI, p. 143-160.

Kutz, F.W., Yobs, A.R., and Yang, S.C., 1976, National pesticide monitoring programs, *in* Lee, R.E., Jr., ed., *Air pollution from pesticides and agricultural processes*: CRC Press, Cleveland, OH, p. 95-136.

Kwok, E.S.C., Atkinson, R., and Arey, J., 1992, Gas-phase atmospheric chemistry of selected thiocarbamates: *Environ. Sci. Technol.*, v. 26, no. 9, p. 1798-1807.

Lane, D.A., Johnson, N.D., Barton, S.C., Thomas, G.H.S., and Schroeder, W.H., 1988, Development and evaluation of a novel gas and particle sampler for semivolatile chlorinated organic compounds in ambient air: *Environ. Sci. Technol.*, v. 22, no. 8, p. 941-947.

Lane, D.A., Schroeder, W.H., and Johnson, N.D., 1992, On the spacial and temporal variations in atmospheric concentrations of hexachlorobenzene and hexachlorocyclohexane isomers at several locations in the province of Ontario, Canada: *Atmospheric Environ.*, v. 26A, no. 1, p. 31-42.

Larsson, P., and Okla, L., 1989, Atmospheric transport of chlorinated hydrocarbons to Sweden in 1985 compared to 1973: *Atmospheric Environ.*, v. 23, no. 8, p. 1699-1711.

Lawson, T.J., and Uk, S., 1979, The influence of wind turbulence, crop characteristics and flying height on the dispersal of aerial sprays: *Atmospheric Environ.*, v. 13, p. 711-715.

Levy, H., II, 1990, Regional and global transport and distribution of trace species released at the earth's surface, *in* Kurtz, D.A., ed., *Long range transport of pesticides*: Lewis Publishers, Inc., Chelsea, MI, p. 83-96.

Lewis, R.G., and Lee, R.E., Jr., 1976, Air pollution from pesticides: Sources, occurrence, and dispersion, *in* Lee, R.E., Jr., ed., *Air pollution from pesticides*: CRC Press, Cleveland, OH, p. 5-50.

Ligocki, M.P., Leuenberger, C., and Pankow, J.F., 1985a, Trace organic compounds in rain, II. Gas scavenging of neutral organic compounds: *Atmospheric Environ.*, v. 19, no. 10, p. 1609-1617.

Ligocki, M.P., Leuenberger, C., and Pankow, J.F., 1985b, Trace organic compounds in rain, III. Particle scavenging of neutral organic compounds: *Atmospheric Environ.*, v. 19, no. 10, p. 1619-1626.

Ligocki, M.P., and Pankow, J.F., 1989, Measurements of the gas/particle distributions of atmospheric organic compounds: *Environ. Sci. Technol.*, v. 23, no. 1, p. 75-83.

Likens, G.E., Bormann, H., Hedin, L.O., Driscoll, C.T., and Eaton, J.S., 1990, Dry deposition of sulfur--A 23-Year record of the Hubbard Brook Forest ecosystem: *Tellus*, v. 42B, no. 4, p. 319-329.

Lindberg, S.E., and Lovett, G.M., 1985, Field measurements of particle dry deposition rates to foliage and inert surfaces in a forest canopy: *Environ. Sci. Technol.*, v. 19, no. 3, p. 238-244.

Mackay, D., Paterson, S., and Schroeder, W.H., 1986, Model describing the rate of transfer of organic chemicals between atmosphere and water: *Environ. Sci. Technol.*, v. 20, p. 810-813.

Maguire, R.J., 1991, Kinetics of pesticide volatilization from surface water: *J. Agri. Food Chem.*, v. 39, no. 9, p. 1674-1678.

Majewski, M.S., Desjardins, R.L., Rochette, P., Pattey, E., Seiber, J.N., and Glotfelty, D.E., 1993, Field comparison of an eddy accumulation and an aerodynamic-gradient system for measuring pesticide volatilization fluxes: *Environ. Sci. Technol.*, v. 27, no. 1, p. 121-128.

Majewski, M.S., Glotfelty, D.E., Paw U, K.T., and Seiber, J.N., 1990, A field comparison of several methods for measuring pesticide evaporation rates from soil: *Environ. Sci. Technol.*, v. 24, no. 10, p. 1490-1497.

Majewski, M.S., McChesney, M.M., and Seiber, J.N., 1991, A field comparison of two methods for measuring DCPA soil evaporation rates: *Environ. Toxicol. Chem.*, v. 10, no. 3, p. 301-311.

Majewski, M.S., McChesney, M.M., Woodrow, J.W., Pruger, J., and Seiber, J.N., 1995, Aerodynamic measurements of methyl bromide volatilization from tarped and nontarped fields: *J. Environ. Qual.*, v. 24, no. 4.

Maybank, J., Yoshida, K., and Grover, R., 1978, Spray drift from agricultural pesticide applications: *J. Air Poll. Control Assoc.*, v. 28, no. 10, p. 1009-1014.

McConnell, L.L., Cotham, W.E., and Bidleman, T.F., 1993, Gas exchange of hexachlorocyclohexane in the Great Lakes: *Environ. Sci. Technol.*, v. 27, no. 7, p. 1304-1311.

McDow, S.R., and Huntzicker, J.J., 1990, Vapor adsorption artifact in the sampling of organic aerosol--Face velocity effects: *Atmospheric Environ.*, v. 24A, no. 10, p. 2563-2571.

Medved, L.I., 1975, Circulation of pesticides in the biosphere: *Pure and Applied Chemistry*, v. 42, p. 119-128.

Menges, R.M., 1964, Influence of wind on performance of preemergence herbicides: *Weeds*, v. 12, no. 3, p. 236-237.

Middleton, J.T., 1965, The presence, persistence, and removal of pesticides in air, *in* Chichester, C.O., ed., *Research in pesticides*: Academic Press, New York, p. 191-197.

Moilanen, K.W., Crosby, D.G., Woodrow, J.E., and Seiber, J.N., 1977, Vapor-phase photocomposition of pesticides in the presence of oxidant: Presented at 173rd National Meeting of the American Chemical Society, New Orleans, LA, March 21, 1977.

Monger, K., and Miller, G.C., 1988, Vapor phase photolysis of trifluralin in an outdoor chamber: *Chemosphere*, v. 17, no. 11, p. 2183-2188.

Monteith, J.L., 1973, *Principles of environmental physics*: Elsevier, New York, 241 p.

Montgomery, J.H., 1993, *Agrochemicals desk reference. Environmental data.* 1st ed: Lewis Publishers, Inc., Chelsea, MI, 625 p.

Muir, D.C.G., Grift, N.P., and Strachan, W.M.J., 1990, Herbicides in rainfall in northwest Ontario, 1989: Presented at the 18th Workshop on the Chemistry and Biochemistry of Herbicides, Regina, Saskatchewan, Canada, April 24-25, 1990.

Muir, D.C.G., Segstro, M.D., Welbourn, P.M., Toom, D., Eisenreich, S.J., Macdonald, C.R., and Whelpdale, D.M., 1993, Patterns of accumulation of airborne organochlorine contaminants in lichens from the upper Great Lakes region of Ontario: *Environ. Sci. Technol.*, v. 27, no. 6, p. 1201-1210.

Munson, T.O., 1976, A note on toxaphene in environmental samples from the Chesapeake Bay region: *Bull. Environ. Contam. Toxicol.*, v. 16, no. 4, p. 491-494.

Murphy, T.J., 1984, Atmospheric inputs of chlorinated hydrocarbons to the Great Lakes, *in* Nriagu, J.O., and Simmins, M.S., eds., *Toxic contaminants in the Great Lakes*: John Wiley and Sons, New York, p. 53-79.

Murray, J.A., and Vaughan, L.M., 1970, Measuring pesticide drift at distances to four miles: *J. Appl. Meteorol.*, v. 9, no. 1, p. 79-85.

Nash, R.G., and Hill, B.D., 1990, Modeling pesticide volatilization and soil decline under controlled conditions, *in* Kurtz, D.A., ed., *Long range transport of pesticides*: Lewis Publishers, Inc., Chelsea, MI, p. 17-28.

National Academy of Sciences and National Academy of Engineering, 1973 [1974], Water quality criteria, 1972: U.S. Environmental Protection Agency, EPA R3-73-033, 594 p.

Nations, B.K., and Hallberg, G.R., 1992, Pesticides in Iowa precipitation: *J. Environ. Qual.*, v. 21, p. 486-492.

Nicholson, K.W., 1988a, The dry deposition of small particles--A review of experimental measurement: *Atmospheric Environ.*, v. 22, no. 12, p. 2653-2666.

-----1988b, A review of particle resuspension: *Atmospheric Environ.*, v. 22, no. 12, p. 2639-2651.

Nigg, H.N., Henry, J.A., and Stamper, J.H., 1983, Regional behavior of pesticide residues in the United States: *Residue Reviews*, v. 85, p. 257-276.

Noll, K.E., and Fang, K.Y.P., 1989, Development of a dry deposition model for atmospheric coarse particles: *Atmospheric Environ.*, v. 233, no. 3, p. 585-594.

Nordby, A., and Skuterud, R., 1975, The effects of boom height, working pressure and wind speed on spray drift: *Weed Res.*, v. 14, p. 385-395.

Nowell, L.H., and Resek, E.A., 1994, National standards and guidelines for pesticides in water, sediment, and aquatic organisms: Application to water-quality assessments: *Rev. Environ. Contam. Toxicol.*, v. 140, 221 p.

Occupational Safety and Health Administration, 1989, Occupational Safety and Health Administration limits for air concentrations: *Federal Register*, v. 54, no. 12, p. 2923-2958.

Oshima, R.J., Mischke, T.M., and Gallavan, R.E., 1980, Drift studies from the aerial application of DEF and folex in Fresno and Merced counties, California, 1979: California Department of Food and Agriculture, EH 80-04, 50 p.

Oshima, R.J., Neher, L.A., Mischke, T.M., Weaver, D.J., and Leifson, O.S., 1982, A characterization of sequential aerial malathion applications in the Santa Clara Valley of California, 1981: California Department of Food and Agriculture, EH 82-01, 137 p.

Oudiz, D., and Klein, A.K., 1988, Evaluation of ethyl parathion as a toxic air contaminant: California Department of Food and Agriculture, EH-88-5, 179 p.

Pacyna, J.M., and Oehme, M., 1988, Long-range transport of some organic compounds to the Norwegian arctic: *Atmospheric Environ.*, v. 22, no. 2, p. 243-257.

Pankow, J.F., 1987, Review and comparative analysis of the theories on partitioning between the gas and aerosol particulate phases in the atmosphere: *Atmospheric Environ.*, v. 21, no. 11, p. 2275-2283.

-----1988, The calculated effects of non-exchangeable material on the gas-particle distribution of organic compounds: *Atmospheric Environ.*, v. 22, no. 7, p. 1405-1409.

-----1991, Common y-intercept and single compound regression of gas-particle partitioning data vs. 1/T: *Atmospheric Environ.*, v. 25A, no. 10, p. 2229-2239.

-----1992, Application of common y-intercept regression parameters for log Kp vs. 1/T for predicting gas-particle partitioning in the urban environment: *Atmospheric Environ.*, v. 26A, no. 14, p. 2489-2497.

-----1994, An absorbtion model of gas/particle partitioning of organic compounds in the atmosphere: *Atmospheric Environ.*, v. 28, no. 2, p. 185-188.

Pankow, J.F., and Bidleman, T.F., 1991, Effects of temperature, TSP and percent non-exchangeable material in determining the gas-particle partitioning of organic compounds: *Atmospheric Environ.*, v. 25A, no. 10, p. 2241-2249.

Pankow, J.F., Isabelle, L.M., and Asher, W.E., 1984, Trace organic compounds in rain. 1. Sampler design and analysis by adsorption/thermal desorption (ATD): *Environ. Sci. Technol.*, v. 18, no. 5, p. 310-318.

Pankow, J.F., Isabelle, L.M., Asher, W.E., Kristensen, T.J., and Peterson, M.E., 1983, Organic compounds in Los Angeles and Portland rain--Identities, concentrations, and operative scavenging mechanisms, *in* Pruppacher, H.R., Semonin, R.G., and Slinn, W.G.N., eds., *Precipitation scavenging, dry deposition, and resuspension*: Elsevier, New York, v. 1, p. 403-414.

Pankow, J.F., Storey, J.M.E., and Yamasaki, H., 1993, Effects of relative humidity on gas/particle partitioning of semivolatile organic compounds to urban particulate matter: *Environ. Sci. Technol.*, v. 27, no. 10, p. 2220-2226.

Patton, G.W., Hinckley, D.A., Walla, M.D., and Bidleman, T.F., 1989, Airborne organochlorines in the Canadian high Arctic: *Tellus*, v. 41B, no. 3, p. 243-255.

Patton, G.W., Walla, M.D., Bidleman, T.F., and Barrie, L.A., 1991, Polycyclic aromatic and organochlorine compounds on the atmosphere of northern Ellesmere Island, Canada: *J. Geophys. Res.*, v. 96, no. D6, p. 10867-10877.

Paw U, K.T., 1992, Rebound and reentrainment of large particles, *in* Schwartz, S.E., and Slinn, W.G.N., eds., *Precipitation scavenging and atmospheric-surface exchange*: Hemisphere Publishing Corp., Washington, D.C., p. 1153-1163.

Payne, N.J., 1992, Off-target glyphosate from aerial silvicultural applications, and buffer zones requires around sensitive areas: *Pest. Sci.*, v. 34, p. 1-8.

Payne, N.J., Sundaram, K.M., and Helson, B.V., 1991, Airborne permethrin and off-target deposits from an aerial ultra-low-volume silviculture spray: *Crop Protection*, v. 10, p. 357-362.

Payne, N.J., and Thompson, D.G., 1992, Off-target glyphosate deposits from aerial silvicultural applications under various meteorological conditions: *Pest. Sci.*, v. 34, p. 53-59.

Peakall, D.B., 1976, DDT in rainwater in New York following application in the Pacific Northwest: *Atmospheric Environ.*, v. 10, p. 899-900.

Peterle, T.J., 1969, DDT in Antarctic snow: *Nature*, v. 224, p. 620.

Pimentel, D., and Levitan, L., 1986, Pesticides--Amounts applied and amounts reaching pests: *BioScience*, v. 36, no. 2, p. 86-91.

Que Hee, S.S., Sutherland, R.G., and Vetter, M., 1975, GLC analysis of 2,4-D concentrations in air samples from central Saskatchewan in 1972: *Environ. Sci. Technol.*, v. 9, no. 1, p. 62-66.

Rahn, K.A., 1981, Relative importance of North America and Eurasia as sources of arctic aerosol: *Atmospheric Environ.*, v. 15, no. 8, p. 1447-1455.

Rapaport, R.A., and Eisenreich, S.J., 1986, Atmospheric deposition of toxaphene to eastern North America derived from peat accumulation: *Atmospheric Environ.*, v. 20, no. 12, p. 2367-2379.

Rapaport, R.A., and Eisenreich, S.J., 1988, Historical atmospheric inputs of high molecular weight chlorinated hydrocarbons to eastern North America: *Environ. Sci. Technol.*, v. 22, no. 8, p. 931-941.

Rapaport, R.A., Urban, N.R., Capel, P.D., Baker, J.E., Looney, B.B., Eisenreich, S.J., and Gorham, E., 1985, "New" DDT inputs to North America--Atmospheric deposition: *Chemosphere*, v. 14, no. 9, p. 1167-1173.

Reisinger, L.M., and Robinson, E., 1976, Long-distance transport of 2,4-D: *J. Appl. Meteorol.*, v. 15, p. 836-845.

Rice, C.P., and Evans, M.S., 1984, Toxaphene in the Great Lakes, *in* Nriagu, J.O., and Simmons, M.S., eds., *Toxic contaminants in the Great Lakes*: John Wiley and Sons, Inc., New York, p. 163-194.

Rice, C.P., Samson, P.J., and Noguchi, G.E., 1986, Atmospheric transport of toxaphene to Lake Michigan: *Environ. Sci. Technol.*, v. 20, no. 11, p. 1109-1116.

Richards, R.P., Kramer, J.W., Baker, D.B., and Krieger, K.A., 1987, Pesticides in rainwater in the northeastern United States: *Nature*, v. 327, no. 14 May, p. 129-131.

Riley, C.M., and Wiesmer, C.J., 1989, Off-target deposition and drift of aerially applied agricultural sprays: *Pest. Sci.*, v. 26, p. 159-166.

Riley, C.M., Wiesmer, C.J., and Ecobichon, D.J., 1989, Measurement of aminocarb in long-distance drift following aerial application to forests: *Bull. Environ. Contam. Toxicol.*, v. 42, p. 37-44.

Risebrough, R.W., 1990, Beyond long-range transport--A model of a global gas chromatographic system, *in* Kurtz, D.A., ed., *Long range transport of pesticides*: Lewis Publishers, Inc., Chelsea, MI, p. 417-426.

Roberts, T.R., and Stoydin, G., 1976, The degradation of (Z)- and (E)-1,3-dichloropropenes and 1,2-dichloropropane in soil: *Pest. Sci.*, v. 7, p. 325-335.

Rosenberg, N.J., Blad, B.L., and Verma, S.B., 1983, *Microclimate* (2d ed): John Wiley and Sons, New York, 495 p.

Ross, L.J., Nicosia, S., Hefner, K.L., Gonzalez, D.A., McChesney, M.M., and Seiber, J.N., 1989, Volatilization, off-site deposition, dissipation, and leaching of DCPA in the field: California Department of Food and Agriculture, EH-89-2, 38 p.

Ross, L.J., Nicosia, S., McChesney, M.M., Hefner, K.L., Gonzalez, D.A., and Seiber, J.N., 1990, Volatilization, off-site deposition, and dissipation of DCPA in the field: *J. Environ. Qual.*, v. 19, no. 4, p. 715-722.

Ross, L.J., and Sava, R.J., 1986, Fate of thiobencarb and molinate in rice fields: *J. Environ. Qual.*, v. 15, no. 3, p. 220-225.

Ross, L.J., Weaver, D.J., Sava, R.J., and Saini, N., 1986, Off-target drift of MCPA--"Real-time" air sampling: California Department of Food and Agriculture, April, 1986, 25 p.

Sava, R.J., 1985, Monterey county residential air monitoring: California Department of Food and Agriculture, Report 85-7, 14 p.

Schmitt, C.J., Ribick, M.A., Ludke, J.L., and May, T.W., 1983, National pesticide monitoring program: Organochlorine residues in freshwater fish, 1976-79: U.S. Department of the Interior, Fish and Wildlife Service, Resource publication 152, 62 p.

Schmitt, C.J., Zajicek, J.L., and Peterman, P.H., 1990, National contamination biomonitoring program: Residues of organochlorine chemicals in U.S. freshwater fish, 1976-1984: *Arch. Environ. Contam. Toxicol.*, v. 19, p. 748-781.

Schomburg, C.J., Glotfelty, D.E., and Seiber, J.N., 1991, Pesticide occurrence and distribution in fog collected near Monterey, California: *Environ. Sci. Technol.*, v. 25, no. 1, p. 155-160.

Schrimpff, E., 1984, Organic micropollutants' balances in watersheds of northeastern Bavaria: *Fresenius Zeitschrift für Analytische Chemie*, v. 319, p. 147-151.

Schroeder, W.H., and Lane, D.A., 1988, The fate of toxic airborne pollutants: *Environ. Sci. Technol.*, v. 22, no. 3, p. 240-246.

Schwack, W., and Bourgeois, B., 1989, Fungicides and photochemistry: Iprodione, procymidone, vinclozolin 1. Photodehalogenation: *Z. Lebensm Unters Forsch*, v. 188, p. 346-347.

Science Advisory Board, 1980, Assessment of air-borne contaminants in the Great Lakes ecosystem. An appendix to the annual report to the international joint commission: Windsor, Ontario.

Segawa, R., Weaver, D., Sava, R., Marade, J., and Oshima, R., 1983, Helicopter spray trial with Sevin XLR (carbaryl), Tulare county 1982: California Department of Food and Agriculture, Environmental Hazards Assessment Program, 40 p.

Segawa, R.T., Sitts, J.A., White, J.H., Marade, S.J., and Powell, S.J., 1991, Environmental monitoring of malathion aerial applications used to eradicate Mediterranean fruit flies in Southern California, 1990: California Department of Food and Agriculture, EH-91-3, 49 p.

Sehmel, G.A., 1980, Particle and gas dry deposition--A review: *Atmospheric Environ.*, v. 14, no. 9, p. 983-1011.

Seiber, J.N., Madden, S.C., McChesney, M.M., and Winterlin, W.L., 1979, Toxaphene dissipation from treated cotton field environments--Component residual behavior on leaves and in air, soil, and sediments determined by capillary gas chromatography: *J. Agri. Food Chem.*, v. 27, no. 2, p. 284-291.

Seiber, J.N., and McChesney, M.M., 1986, Measurement and computer model simulation of the volatilization flux of molinate and methyl parathion from a flooded rice field: California Department of Food and Agriculture, December, 1986, 74 p.

Seiber, J.N., McChesney, M.M., Sanders, P.F., and Woodrow, J.E., 1986, Models for assessing the volatilization of herbicides applied to flooded rice fields: *Chemosphere*, v. 15, no. 2, p. 127-138.

Seiber, J.N., McChesney, M.M., and Woodrow, J.E., 1989, Airborne residues resulting from use of methyl parathion, molinate and thiobencarb on rice in the Sacramento Valley, California: *Environ. Toxicol. Chem.*, v. 8, p. 577-588.

Seiber, J.N., Wilson, B.W., and McChesney, M.M., 1993, Air and fog deposition residues of four organophosphorous insecticides used on dormant orchards in the San Joaquin Valley, California: *Environ. Sci. Technol.*, v. 27, no. 10, p. 2236-2243.

Seinfeld, J.H., 1986, *Atmospheric chemistry and physics of air pollution*: John Wiley and Sons, Inc., New York, p. 738.

Shaffer, K.A., Fritton, D.D., and Baker, D.E., 1979, Drainage water sampling in a wet, dual-pore soil system: *J. Environ. Qual.*, v. 8, p. 241-246.

Shaw, G.E., 1989, Aerosol transport from sources to ice sheets, *in* Oeschger, H., and Langway, C.C., Jr., eds., *The environmental record in glaciers and ice sheets*: John Wiley and Sons, New York, p. 13-27.

Shulters, M.V., Oltman, R.N., and Grabble, R.R., 1987, Pesticides in rainfall in samples collected at Fresno, California, December 1981 through March 1983. Selected Papers in the Hydrological Sciences 1987: U.S. Geological Survey Water-Supply Paper 2330, p. 35-40.

Sickles, J.E., Hodson, L.L., McClenny, W.A., Paur, R.J., Ellestad, T.G., Mulik, J.D., Anlauf, K.G., Wiebe, H.A., Mackay, G.I., Schiff, H.I., and Bubacz, D.K., 1990, Field comparison of methods for the measurement of gaseous and particulate contributors to acidic dry deposition: *Atmospheric Environ.*, v. 24A, no. 1, p. 155-165.

Sievering, H., 1986, Gradient measurements of sulfur and soil mass dry deposition rates under clean air and high wind speed conditions: *Atmospheric Environ.*, v. 20, p. 341-345.

Simmleit, N., and Herrmann, R., 1987a, The behavior of hydrophobic, organic micropollutants in different karst water systems. I. Transport of micropollutants and contaminant balances during the melting of snow: *J. Water, Air, Soil Poll.*, v. 34, p. 79-95.

-----1987b, The behavior of hydrophobic, organic micropollutants in different karst water systems. II. Filtration capacity of karst systems and pollutant sinks: *J. Water, Air, Soil Poll.*, v. 34, p. 97-109.

Simson, T.W., and Cunningham, R.L., 1982, The occurrence of flow channels in soils: *J. Environ. Qual.*, v. 11, no. 1, p. 29-30.

Sirois, A., and Barrie, L.A., 1988, An estimate of the importance of dry deposition as a pathway of acidic substances from the atmosphere to the biosphere in eastern Canada: *Tellus*, v. 40B, p. 59-80.

Sladen, W.J.L., Menzie, C.M., and Reichel, W.L., 1966, DDT residues in Adelie penguins and a crabeater seal from Antarctica: *Nature*, v. 210, p. 670-673.

Slinn, W.G.N., 1977, Some approximations for the wet and dry removal of particles and gases from the atmosphere: *J. Water, Air, Soil Poll.*, v. 7, p. 513-543.

Slinn, W.G.N., Hasse, L., Hicks, B.B., Hogan, A.W., Lal, D., Liss, P.S., Munnich, K.O., Sehmel, G.A., and Vittori, O., 1978, Some aspects of the transfer of atmospheric trace constituents past the air-sea interface: *Atmospheric Environ.*, v. 12, p. 2055-2087.

Slinn, S.A., and Slinn, W.G.N., 1980, Predictions for particle deposition on natural waters: *Atmospheric Environ.*, v. 14, p. 1013-1016.

Smith, A.E., Grover, R., Cessna, A.J., Shewchuk, S.R., and Hunter, J.H., 1986, Fate of diclofop-methyl after application to a wheat field: *J. Environ. Qual.*, v. 15, no. 3, p. 234-238.

Smith, F.B., and Hunt, R.D., 1978, Meteorological aspects of the transport of pollution over long distances: *Atmospheric Environ.*, v. 12, p. 461-477.

Snipes, C.E., Street, J.E., and Mueller, T.C., 1991, Cotton (*Gossypium hirsutum*) response to simulated triclopyr drift: *Weed Tech.*, v. 5, p. 493-498.

-----1992, Cotton (*Gossypium hirutum*) injury from simulated quinclorac drift: *Weed Sci.*, v. 40, p. 106-109.

Soderquist, C.J., Bowers, J.B., and Crosby, D.G., 1977, Dissipation of molinate in a rice field: *J. Agri. Food Chem.*, v. 25, p. 940-945.

Soderquist, C.J., Crosby, D.G., Moilanen, K.W., Seiber, J.N., and Woodrow, J.E., 1975, Occurrence of trifluralin and its photoproducts in air: *J. Agri. Food Chem.*, v. 23, no. 2, p. 304-309.

Spencer, W.F., 1987, Volatilization of pesticide residues, *in* Biggar, J.W., and Seiber, J.N., eds., *Fate of pesticides in the environment*: Davis, CA, Agricultural Experiment Station, University of California Publication 3320, p. 61-68.

Spencer, W.F., and Cliath, M.M., 1970, Desorption of lindane from soil as related to vapor density: *Soil Sci. Soc. America Proc.*, v. 34, p. 574-578.

-----1973, Pesticide volatilization as related to water loss from soil: *J. Environ. Qual.*, v. 2, no. 2, p. 284-289.

-----1974, Factors affecting vapor loss of trifluralin from soil: *J. Agri. Food Chem.*, v. 22, no. 6, p. 987-991.

Spencer, W.F., Cliath, M.M., and Farmer, W.J., 1969, Vapor density of soil-applied dieldrin as related to soil-water content, temperature, and dieldrin concentration: *Soil Sci. Soc. America Proc.*, v. 33, p. 509-511.

Spencer, W.F., Farmer, W.J., and Cliath, M.M., 1973, Pesticide volatilization: *Residue Rev.*, v. 49, p. 1-47.

Spencer, W.F., Farmer, W.J., and Jury, W.A., 1982, Behavior of organic chemicals at soil, air, water interfaces as related to predicting the transport and volatilization of organic pollutants: *Environ. Toxicol. Chem.*, v. 1, p. 17-26.

Spencer, W.F., Jury, W.A., and Farmer, W.J., 1984, Importance of volatilization as a pathway for pesticide loss from forest soils, *in* Garner, W.Y., and Harvey, J., Jr., eds., *Chemical and biological controls in forestry*: American Chemical Society, Washington, D.C., p. 193-210.

Spillman, J.J., 1984, Evaporation from freely falling droplets: *Aeronautical J.*, v. 42, p. 181-185.

Spitzer, T., and Dannecker, W., 1983, Membrane filters as adsorbents for polynuclear aromatic hydrocarbons during high-volume sampling of air particulate matter: *Anal. Chem.*, v. 55, no. 14, p. 2226-2228.

Stanley, C.W., Barney, J.E., II, Helton, M.R., and Yobs, A.R., 1971, Measurement of atmospheric levels of pesticides: *Environ. Sci. Technol.*, v. 5, no. 5, p. 430-435.

Strachan, W.M.J., 1985, Organic substances in the rainfall of Lake Superior: 1983: *J. Environ. Toxicol. Chem.*, v. 4, p. 677-683.

-----1988, Toxic contaminants in rainfall in Canada: 1984: *Environ. Toxicol. Chem.*, v. 7, p. 871-877.

-----1990, Atmospheric deposition of selected organochlorine compounds in Canada, *in* Kurtz, D.A., ed., *Long range transport of pesticides*: Lewis Publishers, Inc., Chelsea, MI, p. 233-240.

Strachan, W.M.J., and Eisenreich, S.J., 1990, Mass balance accounting of chemicals in the Great Lakes, *in* Kurtz, D.A., ed., *Long range transport of pesticides*: Lewis Publishers, Inc., Chelsea, MI, p. 291-301.

Strachan, W.M.J., and Huneault, H., 1979, Polychlorinated biphenyls and organochlorine pesticides in Great Lakes precipitation: *J. Great Lakes Res.*, v. 5, no. 1, p. 61-68.

-----1984, Automated rain sampler for trace organics: *Environ. Sci. Technol.*, v. 18, no. 2, p. 127-130.

Strachan, W.M.J., Huneault, H., Schertzer, W.M., and Elder, F.C., 1980, Organochlorines in precipitation in the Great Lakes region, *in* Afghan, B.K., and Mackay, D., eds., *Hydrocarbons and halogenated hydrocarbons in the aquatic environment (1980)*: Plenum Publishing Corp., New York, p. 387-396.

Suntio, L.R., Shiu, W.Y., Mackay, D., Seiber, J.N., and Glotfelty, D.E., 1988, Critical review of Henry's law constants for pesticides, *Rev. Environ. Contam. Toxicol.*, v. 103, p. 1-59.

Swap, R., Garstang, M., Greco, G.S., Talbot, R., and Kållberg, P., 1992, Saharan dust in the Amazon Basin: *Tellus*, v. 44B, no. 2, p. 133-149.

Sweet, C.W., 1992, International monitoring of the deposition of airborne toxic substances to the Great Lakes: Presented at the 9th World Clean Air Congress, Montreal, Quebec, Canada, September, 1992.

Tabor, E.C., 1965, Pesticides in urban atmospheres: *J. Air Poll. Control Assoc.*, v. 15, no. 9, p. 415-418.

Tanabe, S., Hidaka, H., and Tatsukawa, R., 1983, PCBs and chlorinated hydrocarbon pesticides in Antarctic atmosphere and hydrosphere: *Chemosphere*, v. 12, no. 2, p. 277-288.

Tanabe, S., Tatsukawa, R., Kawano, M., and Hidaka, H., 1982, Global distribution and atmospheric transport of chlorinated hydrocarbons: HCH (BHC) isomers and DDT compounds in the western Pacific, eastern Indian and Antarctic Oceans: *J. Oceanographical Soc. Jpn.*, v. 38, p. 137-148.

Tarrason, L., and Iversen, T., 1992, The influence of North American anthropogenic sulphur emissions over western Europe: *Tellus*, v. 44B, no. 2, p. 114-132.

Tatsukawa, R., Yamaguchi, Y., Kawano, M., Kannan, N., and Tanabe, S., 1990, Global monitoring of organochlorine insecticides--An 11 year case study (1975-1985) of HCHs and DDTs in the open ocean atmosphere and hydrosphere, *in* Kurtz, D.A., ed., *Long range transport of pesticides*: Lewis Publishers, Inc., Chelsea, MI, p. 127-141.

Taylor, A.W., Glotfelty, D.E., Glass, B.L., Freeman, H.P., and Edwards, W.M., 1976, Volatilization of dieldrin and heptachlor from a maize field: *J. Agri. Food Chem.*, v. 24, no. 3, p. 625-631.

Taylor, A.W., Glotfelty, D.E., Turner, B.C., Silver, R.E., Freeman, H.P., and Weiss, A., 1977, Volatilization of dieldrin and heptachlor residues from field vegetation: *J. Agri. Food Chem.*, v. 25, no. 3, p. 542-547.

Taylor, A.W., and Glotfelty, D.E., 1988, Evaporation from soils and crops, *in* Grover, R., ed., *Environmental chemistry of herbicides.* v. 1: CRC Press, Boca Raton, FL, p. 89-129.

Tennekes, H., 1973, A model for the dynamics of the inversion above a convective boundary layer: *J. Atmospheric Sci.*, v. 30, p. 558-567.

Thibodeaux, L.J., Nadler, K.C., Valsaraj, K.T., and Reible, D.D., 1991, The effect of moisture on volatile organic chemical gas-to-particle partitioning with atmospheric aerosols--Competitive adsorption theory predictions: *Atmospheric Environ.*, v. 25A, no. 8, p. 1649-1656.

Thomas, G.W., and Frye, W.W., 1984, Fertilization and liming, *in* Phillips, R.E., and Phillips, S.H., eds., *No-tillage agriculture:* van Nostrand Reinhold Co., New York, p. 87-124.

Thomas, G.W., and Phillips, R.E., 1979, Consequences of water movement in macropores: *J. Environ. Qual.*, v. 8, no. 2, p. 149-152.

Tsai, W., Cohen, Y., Sakugawa, H., and Kaplan, I.R., 1991, Dynamic partitioning of semivolatile organics in gas/particle/rain phases during rain scavenging: *Environ. Sci. Technol.*, v. 25, no. 12, p. 2012-2023.

Turner, B., Powell, S., Miller, N., and Melvin, J., 1989, A field study of fog and dry deposition as sources of inadvertent pesticide residues on row crops: State of California Department of Food and Agriculture, November, 1989, EH 89-11, 42 p.

Turner, B.C., Glotfelty, D.E., and Taylor, A.W., 1977, Photodieldrin formation and volatilization from grass: *J. Agri. Food Chem.*, v. 25, no. 3, p. 548-550.

Turner, B.C., Glotfelty, D.E., Taylor, A.W., and Watson, D.R., 1978, Volatilization of microencapsulated and conventionally applied chlorpropham in the field: *Agronomy J.*, v. 70, p. 933-937.

Turpin, B.J., Liu, S.P., Podolske, K.S., Gomes, M.S.P., Eisenreich, S.J., and McMurry, P.H., 1993, Design and evaluation of a novel diffusion separator for measuring gas/particle distributions of semivolatile organic compounds: *Environ. Sci. Technol.*, v. 27, no. 12, p. 2441-2449.

U.N. Food and Agriculture Organization, 1978-87, Production Yearbook: Food and Agriculture Organization of the United Nations, v. 31-40.

U.S. Environmental Protection Agency, 1992, Water quality standards; establishment of numeric criteria for priority toxic pollutants; States' compliance; Final rule (12/22/92) ("Toxics Rule"): *Federal Register*, v. 57, no. 246, p. 60848-60923.

-----1994a, Drinking water regulations and health advisories: U.S. Environmental Protection Agency, Washington, D.C., EPA 822-R-94-001, 11 p.

-----1994b, Deposition of air pollutants to the Great Waters; First report to congress: U.S. Environmental Protection Agency, Washington, D.C., EPA 452/R-93-055, 89 p.

Van Vaeck, L., Van Cauwenberghe, K., and Janssens, J., 1984, The gas-particle distribution of organic aerosol constituents--Measurement of the volatilization artifact in hi-vol cascade impactor sampling: *Atmospheric Environ.*, v. 18, p. 417-430.

Voldner, E.C., and Schroeder, W.H., 1989, Modelling of atmospheric transport and deposition of toxaphene into the Great Lakes ecosystem: *Atmospheric Environ.*, v. 23, p. 1949-1989.

Wania, F., Mackay, D., and Shiu, W.Y., 1992, Understanding the transfer of low volatility organic contaminants into Arctic regions: *Presented at the 13th meeting of the Soc. of Environ. Toxicol. Chem.*, Cincinnati, OH, 8-12 November, 1992.

Ware, G.W., Apple, E.J., Cahill, W.P., Gerhardt, P.D., and Frost, K.R., 1969, Pesticide drift. II. Mist blower vs. aerial application of sprays: *J. Econ. Entomol.*, v. 62, no. 4, p. 844-846.

Ware, G.W., Morgan, D.P., Estesen, B.J., and Cahill, W.P., 1974, Establishment of reentry intervals for organophosphate-treated cotton fields based on human data. II. Azodrin, ethyl and methyl parathion: *Arch. Environ. Contam. Toxicol.*, v. 2, no. 2, p. 117-129.

Wauchope, R.D., 1987, Effects of conservation tillage on pesticide loss with water, *in* Logan, T.J., Davidson, J.M., Baker, J.L., and Overcash, M.R., eds., *Effects of conservation tillage on groundwater quality--Nitrates and pesticides*: Lewis Publishers, Inc., Chelsea, MI, p. 205-215.

Wauchope, R.D., Buttler, T.M., Hornsby, A.G., Augustijn-Beckers, P.W.M., and Burt, J.P., 1992, The SCS/ARS/CES pesticide properties database for environmental decision-making: *Rev. Environ. Contam. Toxicol.*, v. *123*, p. 1-157.

Weaver, D.J., Hohnson, W.C., and Oshima, R.J., 1983, Helicopter spray trials with carbaryl (XLR): California Department of Food and Agriculture, Environmental Hazards Assessment Program, Environmental Monitoring and Pest Management Branch, 19 p.

West, I., 1964, Pesticides as contaminants: *Arch. of Environ. Health*, v. 9, no. 11, p. 626-633.

Whang, J.M., Schomburg, C.J., Glotfelty, D.E., and Taylor, A.W., 1993, Volatilization of fonofos, chlorpyrifos, and atrazine from conventional and no-till surface soils in the field: *J. Environ. Qual.*, v. 22, p. 173-180.

Whelpdale, D.M., and Moody, J.L., 1990, Large-scale meteorological regimes and transport processes, *in* Knap, A.H., and Kaiser, M., eds., *The long-range atmospheric transport of natural and contaminant substances*: Kluwer Academic Publishers, Boston, MA, p. 3-36.

White, A.W., Jr., Harper, L.A., Leonard, R.A., and Turnbull, J.W., 1977, Trifluralin volatilization losses from a soybean field: *J. Environ. Qual.*, v. 6, no. 1, p. 105-110.

Whitmore, R.W., Kelly, J.E., and Reading, P.L., 1992, National home and garden pesticide use survey. Executive summary, results, and recommendations. 1: Research Triangle Institute, Final report, RTI/5100/17-01F, 140 p.

Whitmore, R.W., Kelly, J.E. Reading, P.L. Brandt, E., and Harris, T., 1993, National home and garden pesticide use survey, *in* Racke, K.D., and Leslie, A.R., eds., *Pesticides in urban environments*: American Chemical Society, Washington, D.C., p. 18-36.

Whittlestone, S., Robinson, E., and Ryan, S., 1992, Radon at the Mauna Loa observatory-- Transport from distant continents: *Atmospheric Environ.*, v. 26A, no. 2, p. 251-260.

Wienhold, B.J., Sadeghi, A.M., and Gish, T.J., 1993, Effect of starch encapsulation and temperature on volatilization of atrazine and alachlor: *J. Environ. Qual.*, v. 22, p. 162-166.

Willis, G.H., McDowell, L.L., Harper, L.A., Southwick, L.M., and Smith, S., 1983, Seasonal disappearance and volatilization of toxaphene and DDT from a cotton field: *J. Environ. Qual.*, v. 12, no. 1, p. 80-85.

Willis, G.H., McDowell, L.L., Smith, S., Southwick, L.M., and Lemon, E.R., 1980, Toxaphene volatilization from a mature cotton canopy: *Agronomy J.*, v. 72, p. 627-631.

Willis, G.H., Parr, J.F., and Smith, S., 1971, Volatilization of soil-applied DDT and DDD from flooded and nonflooded plots: *Pest. Monitor. J.*, v. 4, no. 4, p. 204-208.

Willis, G.H., Parr, J.F., Smith, S., and Carroll, B.R., 1972, Volatilization of dieldrin from fallow soil as affected by different soil water regimes: *J. Environ. Qual.*, v. 1, no. 2, p. 193-196.

Winer, A.M., and Atkinson, R., 1990, Atmospheric reaction pathways and lifetimes for organophosphorous compounds *in* Kurtz, D.A., ed., *Long range transport of pesticides*: Lewis Publishers, Inc., Chelsea, MI, p. 115-126.

Woodrow, J.E., Crosby, D.G., Mast, T., Moilanen, K.W., and Seiber, J.N., 1978, Rates of transformation of trifluralin and parathion vapors in air: *J. Agri. Food Chem.*, v. 26, p. 1312.

Woodrow, J.E., Seiber, J.N., and Crosby, D.G., 1983, Vapor-phase photochemistry of pesticides: *Residue Rev.*, v. 85, p. 111-125.

Woodrow, J.E., Seiber, J.N., Crosby, D.G., Moilanen, K.W., Soderquist, C.J., and Mourer, C., 1977, Airborne and surface residues of parathion and its conversion products in a treated plum orchard environment: *Arch. Environ. Contam. Toxicol.*, v. 6, p. 175-191.

Worthing, C.R., and Walker, S.B., eds., 1987, *The pesticide manual.* 8th ed.: The British Crop Protection Council, Thornton Heath, United Kingdom, 1081 p.

Wu, T.L., 1981, Atrazine residues in estuarine water and the aerial deposition of atrazine into Rhode River, Maryland: *J. Water, Air, Soil Poll.*, v. 15, p. 173-184.

Wu, Y.L., Davidson, C.I., Lindberg, S.E., and Russell, A.G., 1992, Resuspension of particulate chemical species at forested sites: *Environ. Sci. Technol.*, v. 26, no. 12, p. 2428-2435.

Wyngaard, J.C., 1990, Scaler fluxes in the Planetary Boundary Layer--theory, modeling and measurements: *Boundary-Layer Meteorol.*, v. 50, p. 49-77.

Yamasaki, H., Kuwata, K., and Miyamoto, H., 1982, Effects of ambient temperature on aspects of airborne polycyclic aromatic hydrocarbons: *Environ. Sci. Technol.*, v. 16, no. 4, p. 189-194.

Yates, W.E., and Akesson, N.B., 1973, Reducing pesticide chemical drift, *in* van Valkenburg, W., ed., *Pesticide formulations*: Marcel Dekker, New York, p. 275-341.

Yates, W.E., Akesson, N.B., and Bayer, D.E., 1978, Drift of glyphosate sprays applied with aerial and ground equipment: *Weed Sci.* v. 26, no. 6, p. 597-604.

Yates, W.E., Akesson, N.B., and Coeden, R.E., 1974, Criteria for minimizing drift residues on crops downwind from aerial applications: *Trans. American Soc. Agri. Eng.*, v. 17, no. 4, p. 627-632.

Yeary, R.A., and Leonard, J.A, 1993, Measurement of pesticides in air during application to lawns, trees, and shrubs in urban environments, *in* Racke, K.D., and Leslie, A.R., eds., *Pesticides in urban environments*: American Chemical Society, Washington, D.C., p. 275-281.

Young, D.R., McDermott, D.J., and Heesen, T.C., 1976, Aerial fallout of DDT in Southern California: *Bull. Environ. Contam. Toxicol.*, v. 16, no. 5, p. 604-611.

Zabik, J.M., and Seiber, J.N., 1992, Atmospheric transport of organophosphate pesticides from California's Central Valley to the Sierra Nevada mountains: *J. Environ. Qual.*, v. 22, no. 1, p. 80-90.

Zhang, X., and McMurry, P.H., 1991, Theoretical analysis of evaporative losses of adsorbed or absorbed species during atmospheric aerosol sampling: *Environ. Sci. Technol.*, v. 25, no. 3, p. 456-459.

GLOSSARY OF COMMON AND CHEMICAL NAMES OF PESTICIDES

Common name	CAS No.	Chemical abstracts nomenclature
Alachlor	15972-60-8	2-chloro-N-(2,6-diethylphenyl)-N-(methoxymethyl) acetamide
Aldrin	309-00-2	(1α,4α,4aβ,5α,8α,8aβ)-1,2,3,4,10,10-hexachloro-1,4,4a,5,8,8a-hexahydro-1,4:5,8-dimethanonapthylene
Aminocarb	2032-59-9	4-(dimethylamino)-3-methylphenyl methylcarbamate
Atrazine	1912-24-9	6-chloro-N-ethyl-N'(1-methylethyl)-1,3,5-triazine-2,4-diamine
Azinphos-methyl	86-50-0	O,O-dimethyl S-[(4-oxo-1,2,3-benzotriazin-3(4H)-yl) methyl] phosphorodithioate
Azodrin		see Monocrotophos
Bromacil	314-40-9	5-bromo-6-methyl-3-(1-methylpropyl)-2,4-(1H,3H)-pyrimidinedione
Bromoxinil	1689-84-5	3,5-dibromo-4-hydroxybenzonitrile
Butylate	2008-41-5	S-ethyl bis(2-methylpropyl)carbamothiate
Carbaryl	63-25-2	1-naphthalenyl methylcarbamate
Carbofuran	1563-66-2	2,3-dihydro-2,2-dimethylbenzofuranyl methylcarbamate
Chlordane	57-74-9	1,2,4,5,6,7,8,8-octachloro-2,3,3a,4,7,7a-hexahydro- 4,7-methano-1H-indene
Chlordane, cis-	5103-71-9	(1α,2α,3aα,4β,7β,7aα)-1,2,4,5,6,7,8,8-octachloro-2,3,3a,4,7,7a-hexahydro-4,7-methano-1H-indene
Chlordane, $trans$-	5103-74-2	(1α,2β,3aα,4β,7β,7aα)-1,2,4,5,6,7,8,8-octachloro-2,3,3a,4,7,7a-hexahydro-4,7-methano-1H-indene
Chlorpropham	101-21-3	1-methylethyl(3-chlorophenyl)carbamate
Chlorpyrifos	2921-88-2	O,O-diethyl O-(3,5,6-trichloro-2-pyridinyl) phosphorothioate
Chlorthion	500-28-7	O-(3-chloro-4-nitrophenyl)-O,O-dimethyl phosphorothioate

Common name	CAS No.	Chemical abstracts nomenclature
Cyanazine	21725-46-2	2-[[4-chloro-6-(ethylamino)-1,3,5-triazin-2-yl]amino]-2-methylpropanenitrile
D, 2,4-	94-75-7	(2,4-dichlorophenoxy)acetic acid
Dacthal	1861-32-1	dimethyl 2,3,5,6-tetrachloro-1,4-benzenedicarboxylate
DB, 2,4-	94-82-6	4-(2,4-dichlorophenoxy)butanoic acid
DDD, *p,p'*-	72-54-8	1,1'-(2,2-dichloroethylidene)bis[4-chlorobenzene]
DDD, *o,p'*-	53-19-0	1-chloro-2-[2,2-dichloro-1-(4-chlorophenyl)ethyl]benzene
DDE, *p,p'*-	72-55-9	1,1'-(2,2-dichloroethenylidene)bis[4-chlorobenzene]
DDE, *o,p'*-	3424-82-6	1-chloro-2-[(2,2-dichloro-1-(4-chlorophenyl)ethenyl] benzene
DDT, *p,p'*-	50-29-3	1,1'-(2,2,2-trichloroethylidene)bis[4-chlorobenzene]
DDT, *o,p'*-	789-02-6	1-chloro-2-[(2,2,2-trichloro-1-(4-chlorophenyl)ethyl] benzene]
DEF	78-48-8	*S,S,S*-tributyl phosphorotrithioate
Deltamethrin	52918-63-5	[1R-[1α[I*],3α]]-cyano(3-phenoxyphenyl)methyl3-(2,2-dibromoethenyl)-2,2-dimethylcyclopropanecarboxylate
Diazinon	333-41-5	*O,O*-diethyl *O*-[6-methyl-2-(1-methylethyl)-4-pyrimidinyl] phosphorothioate
Dicamba	1918-00-9	3,6-dichloro-2-methoxybenzoic acid
Dichlorprop		see DP, 2,4-
Dicofol	115-32-2	4-chloro-α-(4-chlorophenyl)-α-(trichloromethyl) benzenemethanol
Diclofop	40843-25-2	(RS)-2-[4-(2,4-dichlorophenoxy)phenoxy]propaonic acid
Diclofop-methyl	51338-27-3	methyl 2-[4-(2,4-dichlorophenoxy)phenoxy]propaonic acid

Common name	CAS No.	Chemical abstracts nomenclature
Dieldrin	60-57-1	(1aα,2β,2aα,3β,6β,6aα,7β,7aα)-3,4,5,6,9,9-hexachloro-1a,2,2a,3,6,6a,7,7a-octahydro-2,7:36dimethanonapth[2,3-b]oxirene
Dimethoate	60-51-5	*O,O*-dimethyl *S*-[2-(methylamino)-2oxoethyl]phosphorodithioate
Disulfoton	298-04-4	*O,O*-diethyl *S*-[2-(ethylthio)ethyl]phosphorodithioate
Disyston		see Disulfoton
DP, 2,4-	120-36-5	(±)-2-(2,4-dichlorophenoxy)proprionic acid
Endosulfan	115-29-7	6,7,8,9,10,10-hexachloro-1,5,5a,6,9,9a-hexahydro-6,9-methano-2,4,3-benzodioxathiepin-3-oxide
Endosulfan I	959-98-8	3α,5aβ,6α,9α,9aβ-6,7,8,9,10,10,hexachloro-1,5,5a,6,9,9a-hexahydro-6,9-methano-2,4,3-benzodioxathiepin-3-oxide
Endosulfan II	33213-65-9	3α,5aα,6β,9β,9aβ-6,7,8,9,10,10,hexachloro-1,5,5a,6,9,9a-hexahydro-6,9-methano-2,4,3-benzodioxathiepin-3-oxide
Endrin	72-20-8	(1aα,2β,2aβ,3α,6α,6aβ,7β,7aα)-3,4,5,6,9,9-hexachloro-1a,2,2a,3,6,6a,7,7a-octahydro-2,7:3,6-dimethanonaphth[2,3-b]oxirene
EPTC	759-94-4	*S*-ethyl dipropylcarbamothioate
Ethion	563-12-2	*S,S'*-methylene bis(*O,O*-diethyl)phosphorodithioate
Ethoprop	13194-48-4	*O*-ethyl *S,S*-dipropylphosphorodithioate
Fenitrothion	122-14-5	*O,O*-dimethyl *O*-(3-methyl-4-nitrophenyl) phosphorothioate
Folex	150-50-5	tributyl phosphorotrithioite
Fonofos	944-22-9	(+)-*O*-ethyl *S*-phenylethylphosphonodithioate
Furadan		see Carbofuran
Glyphosate	1071-83-6	*N*-(phosphonomethyl)glycine
HCB	118-74-1	hexachlorobenzene

Common name	CAS No.	Chemical abstracts nomenclature
HCH, α-	319-84-6	1α,2α,3β,4α,5β,6β-hexachlorocyclohexane
HCH, β-	319-85-7	1α,2β,3α,4β,5α,6β-hexachlorocyclohexane
HCH, γ-	58-89-9	1α,2α,3β,4α,5α,6β-hexachlorocyclohexane
HCH, δ-	319-86-8	1α,2α,3α,4β,5α,6β-hexachlorocyclohexane
HCH, technical	608-73-1	1,2,3,4,5,6-hexachlorocyclohexane
Heptachlor	76-44-8	1,4,5,6,7,8,8-heptachloro-3a,4,7,7a-tetrahydro-4,7-methano-1*H*-indene
Heptachlor epoxide	1024-57-3	2,3,4,5,6,7,8-heptachloro-1a,1b,5,5a,6,6a,-hexahydro-2,5-methano-2*H*-indeno(1,2b)oxirene
Kelthane		see Dicofol
Leptophos	21609-90-5	*O*-(4-bromo-2,5-dichlorophenyl) *O*-methylphenyl phosphonothioate
Linuron	330-55-2	*N'*-(3,4-dichlorophenyl)-*N*-methoxy-*N*-methylurea
Malathion	121-75-5	diethyl [(dimethoxyphosphinothioy)lthio]-butanedioate
MCPA	94-74-6	(4-chloro-2-methylphenoxy)acetic acid
MCPB	94-81-5	4-(4-chloro-2-methylphenoxy)butanoic acid
Mecoprop	93-65-2	(±)-2-(4-chloro-2-methylphenoxy)propanoic acid
Methidathion	950-37-8	[(5-methoxy-2-oxo-1,3,4-thiadiazol-3(2H)-yl)methyl] *O,O*-dimethylphosphorodithioate
Methomyl	16752-77-5	methyl *N*-[[(methylamino)carbonyl]oxy] ethanimidothioate
Methoxychlor	72-43-5	1,1'-(2,2,2-trichloroethylidene)bis[4-methoxybenzene]
Methyl bromide	74-83-9	bromomethane
Methyl parathion	298-00-0	*O,O*-dimethyl *O*-(4-nitrophenyl) phosphorothioate
Methyl trithion	953-17-3	*S*-[[(4-chlorophenyl)thio]methyl] *O,O*-dimethyl phosphorodithioate

Common name	CAS No.	Chemical abstracts nomenclature
Metolachlor	51218-45-2	2-chloro-*N*-(2-ethyl-6-methylphenyl)-*N*-(2-methoxy-1-methylethyl)acetamide
Metribuzin	21087-64-9	4-amino-6-(1,1-dimethylethyl)-3-(methylthio)-1,2,4-triazin-5(4*H*)-one
Mevinphos	7786-34-7	methyl 3[(dimethoxyphosphinyl)oxy]-2-butenoate
Mirex	2385-85-5	1,1a,2,2,3,3a,4,5,5,5a,5b,6-dodecachlorooctahydro-1,3,4-metheno-1*H*-cyclobuta[cd]pentalene
Molinate	2212-67-1	*S*-ethyl hexahydro-1*H*-azepine-1-carbothioate
Monocrotophos	6923-22-4	(E)-dimethyl-1-methyl-3-(methylamino)-3-oxo-1-propenyl phosphate
Nitrapyrin	1929-82-4	2-chloro-6-(trichloromethyl)pyridine
Nonachlor, *cis*-	5103-73-1	1,2,5,3,4,5,6,7,8,8-nonachloro-2,3,3a,4,7,7a-hexahydro-4,7-methano-1*H*-indene (combined nomenclature for *cis*- and *trans*-)
Nonachlor, *trans*-	39765-80-5	see Nonachlor, *cis*-
Oxychlordane	27304-13-8	2,3,4,5,6,6a,7,7-octachloro-1a,1b,5,5a,6,6-hexahydro-2,5-methano-2*H*-indeno(1,2b)oxirene
Parathion	56-38-2	*O,O*-diethyl *O*-(4-nitrophenyl) phosphorothioate
Pendimethalin	40487-42-1	*N*-(1-ethylpropyl)-3,4-dimethyl-2,6-dinitro-benzeneamine
Permethrin	52645-53-1	(3-phenoxyphenyl)methyl 3-(2,2-dichloroethenyl)-2,2-dimethylcyclopropanecarboxylate
Phosdrin		see Mevinphos
Phorate	298-02-2	*O,O*-diethyl *S*-[(ethylthio)methyl] phosphorodithioate
Prometon	1610-18-0	6-methoxy-*N,N'*-bis(1-methylethyl)-1,3,5-triazine-2,4-diamine
Prometryn	7287-19-6	*N,N'*-bis(1-methylethyl)6-(methylthio)-1,3,5-triazine-2,4-diamine

Common name	CAS No.	Chemical abstracts nomenclature
Pronamide	23950-58-5	3,5-dichloro-*N*-(1,1-dimethyl-2-propynyl)benzamide
Propachlor	1918-16-7	2-chloro-*N*-(1-methylethyl)-*N*-phenylacetamide
Propanil	709-98-8	*N*-(3,4-dichlorophenyl)propanamide
Propazine	139-40-2	6-chloro-*N,N*'-bis(1-methylethyl)-1,3,5-triazine-2,4-diamine
Ronnel	299-84-3	*O,O*-dimethyl *O*-(2,4,5-trichlorophenyl)phosphorothioate
Silvex		see TP, 2,4,5-
Simazine	122-34-9	2-chloro-*N,N*'-diethyl-1,3,5-triazine-2,4-diamine
T, 2,4,5-	93-76-5	(2,4,5-trichlorophenoxy)acetic acid
TDE, *p,p*'-		see DDD, *p,p*'-
Terbufos	13071-79-9	*S*-[[(1,1-dimethylethyl)thio]methyl] *O,O*-diethyl phosphorodithioate
Tetradifon	116-29-0	1,2,4-trichloro-5-[(4-chlorophenyl)sulfonyl]benzene
Thiobencarb	28249-77-6	*S*-[(4-chlorophenyl)methyl] diethylcarbamothioate
Toxaphene	8001-35-2	polychlorinated camphenes
TP, 2,4,5-	93-72-1	(±)-2-(2,4,5-trichlorophenoxy)propanoic acid
Triallate	2303-17-5	*S*-(2,3,3-trichloro-2-propenyl) bis(1-methylethyl) carbamothioate)
Trifluralin	1582-09-8	2,6-dinitro-*N,N*-dipropyl-4-(trifluoromethyl) benzenamine
Trithion	786-19-6	*S*-[[(4-chlorophenyl)thio]methyl] *O,O*-diethyl phosphorodithioate
Zineb	12122-67-7	[[1,2-ethanediylbis[carbamodithioato]](2-)]zinc
Ziram	137-30-4	(T-4)-bis(dimethyldithiocarbamato-*S,S*')zinc

Index

A

Abamectin, 84

Acephate, 76, 82, 148

Acetanilides, 20, 164. *See also* specific pesticides; specific types
  agricultural use of, 84-85, 106
  home use of, 84-85
  national studies of, 78, 81, 102-108
  national use of, 84-85, 88, 106, 108, 112, 114

Acid-catalyzed hydrolysis, 150

Acifluorfen, 85, 149

Adsorption gains, 153

Aerial spraying, 115, 116, 136

Aerosols, 5, 126, 129

Agricultural management practices, 136-139. *See also* specific types

Agricultural regions, 88

Agricultural use, 4, 77, 82-87, 88, 164. *See also* under specific pesticides
  of acetanilides, 84-85, 106
  of fungicides, 86-87
  of herbicides, 85-86
  of insecticides, 83-84
  of organochlorines, 82
  of organophosphorus compounds, 82-83, 100
  regional, 109
  of triazines, 84-85, 106

Agroecosystem chambers, 138

Air, 5, 20, 77, 80, 88, 90, 101, 114, 141, 163, 166
  aquatic life and, 157, 158
  human health and, 157, 158
  national studies of detection in, 81
  observed concentrations in, 159-160, 161
  organophosphorus compounds in, 99
  quality criteria for, 159-160
  seasonal trends in detection in, 135

Aircraft applications, 115, 116, 136, 138

Air quality criteria, 159-160

Aitken particles, 121

Alabama, 67, 68, 99, 109

Alachlor, 18
  agricultural use of, 84, 106
  chemical name of, 186
  chemical properties of, 146, 148
  climate and, 143
  distribution studies of, 37
  home use of, 84
  local studies of, 54, 57, 58, 62
  maximum concentration levels for, 158, 159, 161
  multistate studies of, 72, 75
  national studies of, 72, 75, 78, 79, 102
  national use of, 84, 106, 108, 148
  physical properties of, 146, 148
  ranking of national use of, 148
  seasonal trends in detection of, 131, 132, 133
  state studies of, 54, 57, 58, 62
  urban use of, 140
  volatilization of, 120, 138
  water quality criteria for, 159
  wet deposition of, 156

Alanap, 86

Alaska, 144

Aldicarb, 83, 148

Aldrin
  agricultural use of, 82
  chemical name of, 186

9 780367 579654